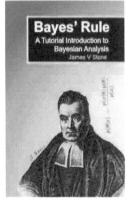

James V Stone is an Honorary Associate Professor at the University of Sheffield, UK.

The Fourier Transform

A Tutorial Introduction

James V Stone

Title: The Fourier Transform
A Tutorial Introduction
Author: James V Stone

©2021 Sebtel Press

First Edition, 2021.
Typeset in LaTeX$\partial 2_\varepsilon$.
First printing.

ISBN 9781916279148

For Bob.

Life welds brothers together
So they both tear at the seam
When one dies.

Contents

Fourier's theorem is not only one of the most beautiful results of modern analysis, but it may be said to furnish an indispensable instrument in the treatment of nearly every recondite question in modern physics.

Lord Kelvin, 1867.

Preface

This book is intended to provide an account of Fourier transforms that is both informal and mathematically rigorous. However, there are no formal proofs in this book, because the main objective is to enable readers to understand the basics of Fourier transforms without having to plough through (or skip over) pages of algebra. Of course, such proofs are necessary, but they can easily be found in more conventional books on Fourier analysis. Additionally, I have included as many diagrams as possible because most readers can understand mathematics in terms of geometry (i.e. visually).

The Fourier transform played a pivotal role in transforming the analogue world of the early 20th century into the digital world of the 21st century. Accordingly, the motivation for writing this book was to provide a firm foundation for understanding Fourier analysis, which underpins fundamental aspects of modern scientific research (e.g. crystallography, CAT scans) as well as our everyday experience of digital communication (e.g. satellite TV, internet video calls).

Who Should Read This Book? The material in this book should be accessible to anyone with an understanding of basic mathematics. The tutorial style adopted ensures that readers who are prepared to put in the effort will be rewarded with a solid grasp of the fundamentals of Fourier transforms.

Online Computer Code. Simple demonstrations of the Fourier transform in Python and Matlab computer code can be downloaded from https://github.com/jgvfwstone/Fourier.

Corrections. Please send any corrections to j.v.stone@sheffield.ac.uk. A complete list of corrections is available on the book web site at https://jim-stone.staff.shef.ac.uk/Fourier.

Acknowledgements. Thanks to Sebastian Stone for singing the single note in Figure 1.8. Thanks to Nikki Hunkin, John Frisby, Steve Snow, and especially Amy Skelt for invaluable feedback. Finally, thanks to Alice Yew for meticulous copyediting and proofreading.

James V Stone.
Sheffield, England, 2021.

Chapter 1

Waves

1.1. Why Fourier Transforms Matter

Everyone is familiar with waves of water from a trip to the seaside, a day on a boat, or stirring a cup of coffee. But waves are not restricted to water (or coffee). They also exist as ripples in the fabric of space, as electromagnetic waves that transfer energy from the Sun to the Earth, and as vibrations in air that deliver sound to our ears. More generally, quantum mechanics has established that even solid matter behaves as if it consists of waves. Accordingly, it is essential to have mathematical tools for analysing the behaviour of waves, whether they are waves of water, electromagnetic radiation, sound, or solid matter. The single most important of these tools is the *Fourier transform*.

The essential, and remarkable, fact that underpins the Fourier transform is that any complicated wave can be separated into *sinusoidal waves* — well, almost any complicated wave; but the exceptions rarely occur in nature. More formally, the Fourier transform is a method of decomposing any complicated wave or *signal* into a set of unique sinusoidal components, as in Figure 1.1a. This decomposition allows each sinusoid to be examined, deleted, attenuated, or amplified. The signal can be any physical quantity, such as sound or electrical current; it can also be an image, a temporal sequence of images, or even a three-dimensional image, as in medical CAT (computerized axial tomography) scans. Applications include live-streaming, data compression, Fraunhofer diffraction, and crystallography. Surprisingly, the Fourier transform also plays a pivotal role in Heisenberg's uncertainty principle, which is a cornerstone of quantum mechanics.

1.2. A Sketch of Fourier Analysis

Tapping a wine glass produces a note at a particular *resonant frequency*. Alternatively, a continuous note at the resonant frequency can be obtained by running a moistened finger around the rim of the glass, as shown in Figure 1.1b. If that note is recorded and played back at high volume, the wine glass will vibrate or *resonate* in sympathy; and if the volume is sufficiently high then the glass will shatter. Even if the resonant frequency of a glass is added to music, the glass still vibrates according to how much of that frequency was added. In other words, the glass 'picks out' its resonant frequency from the sound around it.

What is true of one wine glass is also true of any other glass. Large glasses resonate at low frequencies, whereas small glasses resonate at high frequencies. Alternatively, the resonant frequency of a glass can be altered by partially filling it with water. An array of glasses with different resonant frequencies forms a *glass harmonica*, shown in Figure 1.1b, which has been used as a musical instrument since ancient times.

Just as each glass in a harmonica produces a different note, so each glass 'picks out' one note from the surrounding sound. The smallest glass picks out high frequencies, and the largest glass picks out low frequencies. For example, a violin concerto will induce each glass to vibrate according to the amount of its resonant frequency present in the music. Thus, the harmonica effectively separates the music into its constituent frequencies, and the amplitude of the vibrations induced in each glass indicates the *amplitude* of each frequency in the music.

If a small segment of the music consists of just three notes, as in Figure 1.1a, then three glasses would vibrate at the sinusoidal frequencies shown in *i*, *ii* and *iii*. The heights of these three sinusoids collectively define the *amplitude spectrum* of the music, whereas the left–right positions or *phases* of the sinusoids collectively define the *phase spectrum* of the music. Together, the amplitude spectrum and the phase spectrum constitute the *Fourier transform* of the music. Thus, for all practical purposes, the glass harmonica is an analogue computer that implements a Fourier transform of the surrounding sound.

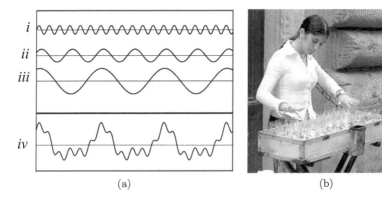

(a) (b)

Figure 1.1: (a) The complicated signal in *iv* is the sum of corresponding points in the sinusoidal curves in *i*, *ii* and *iii*. The Fourier transform decomposes the signal in *iv* into the sinusoids in *i*, *ii* and *iii*. (b) Glass harmonica: when a moistened finger is moved around the rim of a glass, the glass emits a note at its resonant frequency. Photograph by A Pingstone.

1.3. Making Waves

Waves are the basic elements of Fourier transforms, so we need to define exactly what we mean by a wave. Consider a clock that runs backwards, where the initial position of the clock hand is at 3, in the direction of the x-axis, and the angle between the clock hand and the x-axis is θ (theta). The z-axis is at $90°$ to the clock face, so as the clock moves along the z-axis the hand traces out a helix, as shown in Figure 1.2a.

When the helix is viewed from the side, this reveals a sine function, as shown in Figure 1.3. If the length of the hand is A then $\sin\theta = y/A$, so the height of the point at the tip of the hand is

$$y \quad = \quad A\sin\theta, \tag{1.1}$$

as shown in Figures 1.2b. Similarly, when the helix is viewed from above, this reveals a cosine function, as shown in Figure 1.3. If we drop a vertical line from tip of the hand onto the x-axis then it lands at a distance of

$$x \quad = \quad A\cos\theta \tag{1.2}$$

from the centre of the clock face, as shown in Figures 1.2b. The sine and cosine functions are also shown in Figures and 1.4a and and 1.4b (respectively). The distance between consecutive peaks of the sine and cosine functions is the *wavelength* λ (lambda), as shown in Figure 1.3.

Radians. It is more convenient to measure angles in *radians* rather than degrees. A simple way to think about radians is to note that the

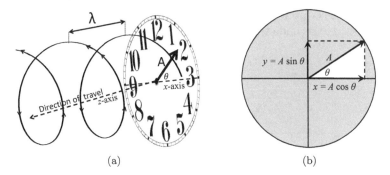

(a)   (b)

Figure 1.2: (a) A clock hand with length A rotating anti-clockwise on a moving clock traces out a helix, with wavelength λ m. The hand makes an angle θ with the horizontal x-axis. If the hand completes ν rotations per second (one rotation is 2π radians) then it has an angular frequency of $\omega = 2\pi\nu$ radians per second (rad/s). Clock face by Karen Watson.
(b) Lengths of the projections of the clock hand onto the x- and y-axes.

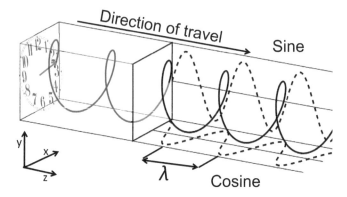

Figure 1.3: Helix generated by the (anti-clockwise rotating) hand of a moving clock. When projected onto a vertical plane this helix produces a sine wave, and when projected onto a horizontal plane it produces a cosine wave.

360 degrees in a circle correspond to 2π radians, so one radian is

$$\frac{360°}{2\pi} \approx 57.3°. \tag{1.3}$$

Temporal Frequency. If the hand rotates through one complete rotation or *cycle* in T seconds then its *temporal frequency* is

$$\nu = 1/T \text{ cycles per second, or hertz (Hz)}, \tag{1.4}$$

where ν is the Greek letter *nu* (pronounced *new*). Temporal frequency is commonly referred to simply as *frequency*.

Angular Frequency. If the hand rotates through one cycle, i.e. 2π radians, in T seconds then its *angular frequency* is

$$\omega = 2\pi/T \text{ radians per second (rad/s)}, \tag{1.5}$$

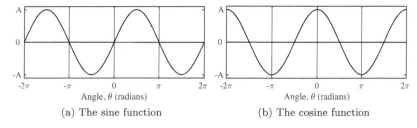

(a) The sine function   (b) The cosine function

Figure 1.4: (a) The sine function $A\sin\theta$ plotted for two complete cycles over $[-2\pi, 2\pi]$ radians. (b) The cosine function $A\cos\theta$ plotted for two complete cycles. Both functions have amplitude A and vary between $-A$ and $+A$.

where ω is the Greek letter *omega*. By substituting Equation 1.4 into Equation 1.5 we can see that the temporal frequency ν and angular frequency ω are related by a factor of 2π:

$$\omega = 2\pi\nu. \tag{1.6}$$

Angular Frequency or Temporal Frequency? The decision of whether to represent a wave in terms of temporal frequency ν or angular frequency ω is, to some extent, a matter of taste, and we will be representing frequency as both ν and ω in this book.

Phase and Amplitude. Given that the clock hand rotates at the rate of ω rad/s and began with an angle of $\theta = 0$ radians, it follows that the angle or *phase* at time t is

$$\theta = \omega t \tag{1.7}$$
$$= 2\pi\nu t \text{ radians}, \tag{1.8}$$

so Equation 1.2 becomes

$$x = A\cos\theta \tag{1.9}$$
$$= A\cos(\omega t). \tag{1.10}$$

The cosine function $\cos\theta$ has a minimum value of -1 at $\theta = \pi$ and a maximum value of $+1$ at $\theta = 0$, so it varies between ± 1. Consequently, the function $A\cos\theta$ varies between $\pm A$, where A is the wave's *amplitude*. Of course, analogous observations apply to the sine function.

More generally, angular frequency and frequency can refer to any physical variable. For example, if we replace time with length then angular frequency measures the change in angle per unit length, known as the *wavenumber*, which has units of radians per metre (rad/m), and frequency measures the number of cycles per unit length as *spatial frequency*, which therefore has units of cycles per metre (m^{-1}).

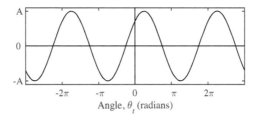

Figure 1.5: A cosine wave with an initial phase of $-\theta' = -\pi/4$ radians ($45°$clockwise from x-axis) has a value of $x_t = A\cos(\theta_t - \theta')$, which varies between $-A$ and A.

1.4. Changing Phases

Consider a sinusoidal wave generated by a clock hand that rotates with an angular frequency of ω rad/s. If the initial angle or phase of the hand is $-\theta'$ (i.e. θ' measured clockwise from the x-axis) then its phase at time t is $\theta_t - \theta'$ where $\theta_t = \omega t$, as shown in Figure 1.5. (We have introduced a t subscript to distinguish the time-varying angle θ_t from the initial angle of $-\theta'$.) If the amplitude of the wave is A then

$$x \;=\; A\cos(\theta_t - \theta'). \tag{1.11}$$

However, it will prove useful to represent a sinusoid using a combination of sine and cosine functions,

$$x \;=\; C\cos\theta_t + D\sin\theta_t \tag{1.12}$$
$$\;=\; C\cos(\omega t) + D\sin(\omega t), \tag{1.13}$$

as shown in Figure 1.6. For example, if $C=0$ then x is a sine, and if $D=0$ then x is a cosine, so Equation 1.12 varies smoothly between a sine and a cosine function as C and D are varied. The constants C and D can be obtained from the known values of A and θ'. Putting Equations 1.11 and 1.12 together, we have

$$A\cos(\theta_t - \theta') \;=\; C\cos\theta_t + D\sin\theta_t. \tag{1.14}$$

Taking the trigonometric identity

$$\cos(\alpha - \beta) \;=\; \cos\alpha\cos\beta + \sin\alpha\sin\beta, \tag{1.15}$$

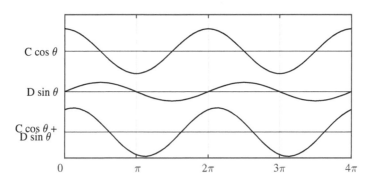

Figure 1.6: Top: cosine function $C\cos\theta_t$ where $C = 1.386$.
Middle: sine function $D\sin\theta_t$ where $D = 0.574$.
Bottom: the sum $C\cos\theta_t + D\sin\theta_t$, which is a sinusoid with amplitude $A = \sqrt{C^2 + D^2} = 1.5$ and initial phase $\theta' = \arctan(D/C) = 0.393$ radians.

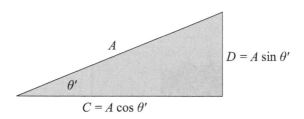

$$D = A \sin \theta'$$

$$C = A \cos \theta'$$

Figure 1.7: A sinusoidal wave with a phase of θ' at time $t = 0$ has an initial height of $C \cos \theta' + D \sin \theta'$, where $C = A \cos \theta'$ and $D = A \sin \theta'$. The *amplitude* is $A = \sqrt{C^2 + D^2}$, and the phase is $\theta' = \arctan(D/C)$.

substituting $\alpha = \theta_t$ and $\beta = \theta'$, and multiplying both sides by A yields

$$A \cos(\theta_t - \theta') \;=\; [A \cos \theta'] \cos \theta_t + [A \sin \theta'] \sin \theta_t. \tag{1.16}$$

Equating the terms in square brackets with the corresponding terms in Equation 1.14 gives

$$C = A \cos \theta' \quad \text{and} \quad D = A \sin \theta', \tag{1.17}$$

as shown in Figure 1.7. Therefore,

$$\cos \theta' = C/A \quad \text{and} \quad \sin \theta' = D/A. \tag{1.18}$$

From the standard trigonometric identity $\sin^2 \theta' + \cos^2 \theta' = 1$, Equations 1.18 imply that

$$(C/A)^2 + (D/A)^2 \;=\; 1, \tag{1.19}$$

and therefore the amplitude is (see Figure 1.7)

$$A \;=\; \sqrt{C^2 + D^2}. \tag{1.20}$$

From Equations 1.18 and Figure 1.7, the initial phase can also be expressed in terms of C and D, as θ' satisfies

$$\tan \theta' \;=\; D/C. \tag{1.21}$$

Since $\theta_t = \omega t$, Equation 1.14 can be written in terms of the angular frequency and time as

$$A \cos[\omega(t - t')] \;=\; C \cos(\omega t) + D \sin(\omega t), \tag{1.22}$$

where $t' = \theta'/\omega$. Finally, note that the pairs of constants (C, D) and (A, θ') refer to a single frequency only.

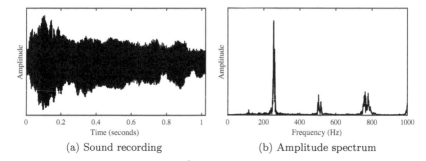

(a) Sound recording   (b) Amplitude spectrum

Figure 1.8: Fourier analysis. (a) Male voice. (b) Amplitude spectrum between 0 and 1000 Hz. Figure generated with computer code in Appendix B.

1.5. A Simple Demonstration

To get some idea of where we are heading, Figure 1.8 gives a simple demonstration of Fourier analysis in action. A sound recording of a single voice is shown in Figure 1.8a. The amplitude of each frequency up to 1000 Hz is shown in Figure 1.8b, which is part of the Fourier transform known as the amplitude spectrum. Notice that this indicates a fundamental frequency of about 270 Hz, with harmonics at around 540 and 810 Hz.

Figure 1.8 can be generated from the computer code in Appendix A (Python) or Appendix B (Matlab), which can be downloaded from https://jim-stone.staff.shef.ac.uk/Fourier.

1.6. Jean-Baptiste Joseph Fourier

Baron Jean-Baptiste Joseph Fourier (1768–1830) was one of several scientists involved in the development of Fourier analysis, the original results of which were published in his book of 1822. However, his methods were unorthodox, and his mathematical derivations did not meet with the approval of the French Academy of Sciences. In awarding Fourier a prize for his work on (what came to be known as) Fourier analysis in 1811, the Academy's begrudging citation was as follows:

> *The novelty of the subject, together with its importance, has decided us to award the prize, while nevertheless observing that the manner in which the author arrives at his equations is not without difficulties, and that his analysis for integrating them still leaves something to be desired both as to generality and even as to rigour.*

It seems likely that the French Academy of Sciences would have been less disparaging if they known that Fourier analysis would underpin many of the key scientific developments of the 20th century.

Chapter 2

Mixing Waves

2.1. Setting the Scene

Our objective is to find the Fourier transform of a temporal signal $f_T(t)$ that lasts for T seconds. However, strictly speaking, Fourier analysis is designed for *periodic* functions, whose graphs are regularly repeating patterns. To meet this requirement, we 'cheat' a little by defining a function $f(t)$ that is many copies of $f_T(t)$ joined end to end, as in Figure 2.1, and perform Fourier analysis on this periodic function $f(t)$.

In the process of finding its Fourier transform, we will effectively re-create the function $f(t)$ from a set of *basis functions* consisting of sine and cosine functions. For a sinusoidal basis function with period T, its angular frequency is

$$\omega_1 \;=\; 2\pi/T \text{ rad/s}, \tag{2.1}$$

as shown in Figure 2.2a. The other sinusoidal basis functions we will use have frequencies that are integer multiples of ω_1, so the nth sinusoidal basis function has a frequency of

$$\omega_n \;=\; n\omega_1 \tag{2.2}$$
$$\;=\; n2\pi/T \text{ rad/s}. \tag{2.3}$$

Thus, ω_1 can be considered the *fundamental frequency*, and successive values of n define the *harmonics* of ω_1 (see Figure 2.2). Notice that, by

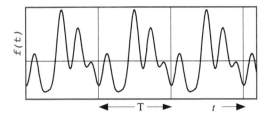

Figure 2.1: A periodic function $f(t)$ with a period of T seconds.

construction, there are exactly n complete cycles of the nth harmonic in T seconds. In principle, we could choose almost any function to demonstrate these ideas, but it will help to use a particular function $f(t)$, as depicted in Figure 2.1.

2.2. Reconstructing Functions from Sinusoids

Suppose we are given a curve $f(t)$ that looks complicated, like the curve in Figure 2.1, which is the same as the curve on the right end panel of Figure 2.3. Despite its apparent complexity, we are assured that the curve has been constructed simply by adding a few sinusoids together. For the moment, our task is to see how such a composite curve can be constructed if we know the phase and amplitude of each component sinusoid. Later, we will discover how to un-mix the composite curve into its constituent sinusoids.

To keep matters simple, we assume that $f(t)$ is the sum of just five sinusoids, where each sinusoid has a frequency that is a multiple of a fundamental angular frequency ω_1 rad/s, as shown in Figure 2.3:

$$\omega_1 = 1 \times \omega_1, \tag{2.4}$$
$$\vdots$$
$$\omega_5 = 5 \times \omega_1. \tag{2.5}$$

If these sinusoids were pure cosines,

$$f_1(t) = \cos(\omega_1 t), \tag{2.6}$$
$$\vdots$$
$$f_5(t) = \cos(\omega_5 t), \tag{2.7}$$

then the composite curve would be

$$f(t) = \cos(\omega_1 t) + \cdots + \cos(\omega_5 t), \tag{2.8}$$

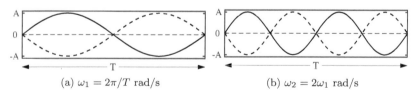

(a) $\omega_1 = 2\pi/T$ rad/s　　　　　　(b) $\omega_2 = 2\omega_1$ rad/s

Figure 2.2: (a) If one complete cycle of 2π radians is swept out in T seconds then the fundamental angular frequency is $\omega_1 = 2\pi/T$ rad/s. (b) The first harmonic is $\omega_2 = 2 \times 2\pi/T = 2\omega_1$ rad/s.

which can be written more succinctly as

$$f(t) = \sum_{n=1}^{5} \cos(\omega_n t). \tag{2.9}$$

To make the situation more realistic, assume that each sinusoid has its own amplitude A_n and its own initial phase $-\theta'_n$ at $t = 0$, so that

$$f_1(t) = A_1 \cos(\omega_1 t - \theta'_1), \tag{2.10}$$
$$\vdots$$
$$f_5(t) = A_5 \cos(\omega_5 t - \theta'_5), \tag{2.11}$$

where each of $f_1(t), \ldots, f_5(t)$ is shown in Figure 2.3. Now Equation 2.9 can be written as

$$f(t) = \sum_{n=1}^{5} A_n \cos(\omega_n t - \theta'_n). \tag{2.12}$$

We can make use of Equation 1.22 to rewrite this as

$$f(t) = \sum_{n=1}^{5} C_n \cos(\omega_n t) + D_n \sin(\omega_n t). \tag{2.13}$$

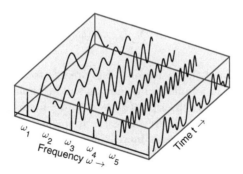

Figure 2.3: The sinusoids in the box have different phases and amplitudes, and the function $f(t)$ on the right end panel is the sum of those sinusoids (the same as in Figure 2.1). The function $f(t)$ repeats every T seconds, so its fundamental frequency is $\omega_1 = 2\pi/T$ rad/s. Fourier analysis decomposes $f(t)$ into the sinusoids inside the box, with frequencies $\omega_1, \ldots, \omega_5$. The amplitudes of the sinusoids are indicated on the frequency axis, and these define the amplitude spectrum.

11

The result of this summation can be seen in Figure 2.3. The height of the curve $f(t)$ at time t is obtained by summing the heights at time t of the five sinusoidal curves (i.e. summing along the frequency axis at time t).

The height A_n of each spike on the frequency axis of Figure 2.3 is the amplitude of the corresponding frequency, so this graph depicts the *amplitude spectrum* of $f(t)$. Similarly, the phases of the sinusoidal components collectively define the *phase spectrum* of $f(t)$ (not shown). Thus, a Fourier transform consists of two spectra, the amplitude spectrum and the phase spectrum.

2.3. Functions with Nonzero Means

Every sinusoid has a mean value of zero, so when many sinusoids are added the result also has a mean of zero. But not all waves in the physical world have a mean of zero; for example, the mean height of a wave in a puddle on the side of Mount Everest is the height of the water surface when there are no waves. We can accommodate such nonzero mean values by adding a constant to the function $f(t)$.

Adding a constant $A_0/2$ to the right-hand side of Equation 2.12 yields

$$f(t) = A_0/2 + \sum_{n=1}^{5} A_n \cos(\omega_n t - \theta'_n). \qquad (2.14)$$

Similarly, adding a constant $C_0/2$ to Equation 2.13 yields

$$f(t) = C_0/2 + \sum_{n=1}^{5} C_n \cos(\omega_n t) + D_n \sin(\omega_n t). \qquad (2.15)$$

Here $C_0 = A_0$, and the factor of $1/2$ is present to allow all coefficients to be calculated in a uniform manner, as will be explained in Chapter 3.

Coefficients with a subscript of zero refer to a frequency of zero, which may sound odd at first. However, as the frequency of a sinusoid is made to decrease, the curve looks increasingly 'flat'. Following this to its logical conclusion, in the limit as $\omega \to 0$, a frequency of zero corresponds to a horizontal line. This implies that the mean value of a function can be expressed as the amplitude of a sinusoid with a frequency of zero.

The Almost-Fourier Transform. At this juncture, we should note that all the information necessary to reconstruct $f(t)$ exactly is implicit in the two sets of phases and amplitudes in Equation 2.14,

$$\begin{aligned} \boldsymbol{A} &= \{A_0/2, A_1, \ldots, A_5\}, \\ \boldsymbol{\theta'} &= \{\theta'_0, \theta'_1, \ldots, \theta'_5\}, \end{aligned} \qquad (2.16)$$

where θ_0' is the phase corresponding to a frequency of zero, and is included purely for symmetry here. This information is also implicit in the sine and cosine coefficients in Equation 2.15,

$$
\begin{aligned}
\boldsymbol{C} &= \{C_0/2, C_1 \dots, C_5\}, \\
\boldsymbol{D} &= \{D_0/2, D_1, \dots, D_5\},
\end{aligned}
\tag{2.17}
$$

where $D_0 = 0$ is also included for symmetry. In fact, both $\{\boldsymbol{A}, \boldsymbol{\theta'}\}$ and $\{\boldsymbol{C}, \boldsymbol{D}\}$ represent a truncated form of the *Fourier transform* of the function $f(t)$. We don't yet know how to find these coefficients — that will be explained later. But first let us see how these ideas can be generalised beyond $n = 5$.

2.4. The Fourier Transform

The complicated composite function $f(t)$ shown in Figure 2.3 was constructed using individual sinusoids with frequencies $\omega_1, \dots, \omega_5$. An obvious way to allow $f(t)$ to become more complicated is to increase the range of frequencies. In fact, almost any function can be constructed if the range of frequencies is allowed to extend out to infinity, so that Equation 2.15 becomes

$$
f(t) = C_0/2 + \sum_{n=1}^{\infty} C_n \cos(\omega_n t) + D_n \sin(\omega_n t)
\tag{2.18}
$$

and Equation 2.14 becomes

$$
f(t) = A_0/2 + \sum_{n=1}^{\infty} A_n \cos(\omega_n t - \theta_n').
\tag{2.19}
$$

By analogy with Equations 2.17, the *sine–cosine Fourier transform* of $f(t)$ consists of the two infinite sets of coefficients $\boldsymbol{C}$ and $\boldsymbol{D}$, where

$$
\begin{aligned}
\boldsymbol{C} &= \{C_0/2, C_1, C_2, C_3, \dots\}, \\
\boldsymbol{D} &= \{D_0/2, D_1, D_2, D_3, \dots\}.
\end{aligned}
\tag{2.20}
$$

Similarly (and equivalently), the *amplitude–phase Fourier transform* comprises the infinite sets

$$
\begin{aligned}
\boldsymbol{A} &= \{A_0/2, A_1, A_2, A_3, \dots\}, \\
\boldsymbol{\theta'} &= \{\theta_0', \theta_1', \theta_2', \theta_3', \dots\}.
\end{aligned}
\tag{2.21}
$$

The Fourier transform described here is also known as the *discrete Fourier transform* (DFT). Remarkably, either the two sets in Equation 2.20 or the two sets in Equation 2.21 provide all the information necessary to reconstruct $f(t)$ exactly. In other words, we can choose to represent the function $f(t)$ either as itself or as its Fourier transform, and the latter, in turn, can be represented either as the sine–cosine Fourier transform in Equation 2.20 or as the phase–amplitude Fourier transform in Equation 2.21. In Chapter 6 we will see how the Fourier transform can also be represented using complex coefficients.

The Inverse Fourier Transform. For later reference, the function $f(t)$ as expressed in Equations 2.18 and 2.19 is called the *inverse Fourier transform* of $\{C, D\}$ and of $\{A, \theta'\}$.

From Sine–Cosine to Amplitude–Phase Transforms. From Equation 1.20, the amplitude at frequency ω_n can be written as

$$A_n \;=\; \sqrt{C_n^2 + D_n^2}, \tag{2.22}$$

which specifies the amount (but not the phase) of each Fourier component in $f(t)$. According to Equation 1.21, the initial phase $-\theta'_n$ of the sinusoidal component at frequency ω_n is given by

$$\tan \theta'_n \;=\; D_n/C_n. \tag{2.23}$$

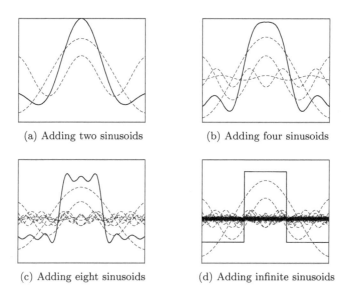

(a) Adding two sinusoids

(b) Adding four sinusoids

(c) Adding eight sinusoids

(d) Adding infinite sinusoids

Figure 2.4: By adding sinusoids with different frequencies, we can approximate a rectangular function. Each sinusoid is shown as a dashed curve, and a weighted sum $f(t)$ of the sinusoids in each panel is drawn as a solid curve. As the number of sinusoids becomes infinite, $f(t)$ becomes the rectangular function shown in (d).

2.5. Approximating a Function

As a tangible demonstration of the Fourier transform, consider Figure 2.4. The solid curve $f(t)$ in each panel is a weighted sum of the sinusoids (dashed curves) in that panel. As the number of sinusoids is increased, $f(t)$ becomes an increasingly good approximation to the rectangular function in (d). The solid curve $f(t)$ in each panel was drawn using Equation 2.19 with $A_0 = 0$, $\theta'_n = 0$, and Fourier coefficients given by $A_n = \sin(\omega_n t)/(\omega_n t)$ (skip to Section 4.4 (p29) for a graph of A_n).

2.6. Summary

The Fourier transform of $f(t)$ can be defined in terms of two equivalent sets of pairs of parameters. Each pair of parameter values determines the amplitude and phase of a sinusoid at one frequency. These pairs of parameters are:

1. the amplitude A and phase θ' of a sinusoid;

2. the amplitude C of a cosine and the amplitude D of a sine, which together define a sinusoid with amplitude $A = \sqrt{C^2 + D^2}$ and phase given by $\theta' = \arctan(D/C)$.

Additionally, each sinusoid can be defined in terms of an *angular frequency* ω, which typically has units of radians per second (rad/s), or in terms of a *frequency* ν, which typically has units of cycles per second, or hertz (Hz).

Of course, knowing that a function can be expressed as the sum of sinusoidal functions is very different from knowing the values of the Fourier coefficients C_n and D_n (or, equivalently, A_n and θ'_n) that allow us to do so. So that will be the topic explored in the next chapter.

Chapter 3

The Parameters of a Single Sinusoid

3.1. Introduction

Our ultimate objective is to find the two sets of coefficients $\boldsymbol{C}$ and $\boldsymbol{D}$ that make up the Fourier transform of a function $f(t)$, which is a periodic function with period T that consists of infinitely many copies of a curve defined on an interval of length T. In preparation, we first solve a simpler version of this problem. Specifically, we will find a single pair of coefficients C_n and D_n for a sinusoidal function $f(\omega_n, t)$ that contains only one sinusoidal frequency ω_n. Using Equations 2.18 and 2.19 we can define

$$f(\omega_n, t) \quad = \quad C_n \cos(\omega_n t) + D_n \sin(\omega_n t) \tag{3.1}$$

$$= \quad A_n \cos(\omega_n t - \theta_n'). \tag{3.2}$$

To avoid the notational overhead associated with $A_0/2$ and $C_0/2$, we assume $A_0 = C_0 = D_0 = 0$ for now, so that

$$f(t) \quad = \quad \sum_{n=1}^{\infty} f(\omega_n, t). \tag{3.3}$$

Notice that $f(\omega_n, t)$ is a bivariate function of frequency ω_n and time t, as shown in Figure 2.3 for five components. However, the bivariate nature of this function is more apparent when using a large number of sinusoidal components, as in Figure 3.1. Now, Equation 3.3 can be recognised as a *marginalisation* of the function $f(\omega_n, t)$. This function is a little unusual in that the time variable t is continuous whereas the frequency variable ω_n is discrete (we will treat frequency as a continuous variable in Section 6.3). Note that if the number of Fourier components in Figure 3.1 is allowed to increase without limit then the curve defined by $f(t)$ will become a perfect rectangle (see front cover or Figure 6.1).

In fact, $f(\omega_n, t)$ can be marginalised to reveal two different quantities. First, marginalisation of $f(\omega_n, t)$ over frequency yields the sum $f(t)$ of all frequencies at one time t (as in Equation 3.3). This marginalisation suits our purposes exactly, because it allows $f(t)$ to be reconstructed.

The second marginalisation averages $f(\omega_n, t)$ over one time period T and calculates, for each frequency ω_n, the integral over all times between $-T/2$ and $T/2$,

$$\int_{t=-T/2}^{T/2} f(\omega_n, t)\, dt. \tag{3.4}$$

The problem is that, by construction, in the interval $[-T/2, T/2]$ there are a complete number of cycles at frequency ω_n, so every sinusoid has a mean value of zero. This is because for every value $y = \cos(\omega t)$ above zero there is a corresponding value $-y = \cos(\pi - \omega t)$ below zero, so the integral in Equation 3.4 equals zero. This problem can be partially overcome by evaluating the integral of $f(\omega_n, t)$ squared, which is always non-negative, and defining

$$f(\omega_n)^2 \;=\; \int_{t=-T/2}^{T/2} f(\omega_n, t)^2\, dt, \tag{3.5}$$

which is T times the *power* at frequency ω_n.

However, $f(\omega_n)^2$ provides only a measure of how much each frequency contributes to $f(t)$ and discards all information regarding the phase of each frequency. But knowing the amount of each component at each frequency is not enough to reconstruct $f(t)$; instead, as is apparent from Equation 2.19, reconstructing $f(t)$ from a sum of sinusoids at different

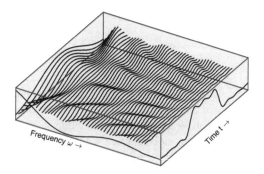

Figure 3.1: Fourier analysis. The curve along the time axis is obtained by summing all the sinusoidal curves in the box. The amplitude A_n of the sinusoid at frequency ω_n is given by the height of the curve drawn along the frequency axis, where $A_n = \sin(\omega_n t)/(\omega_n t)$ (see Figure 4.3a).

frequencies requires knowledge of both the amplitude A_n and the phase θ'_n at each frequency. This problem is considered in the next section.

To reduce notational clutter, from now on the lower and upper limits of integrals with respect to time are assumed to be $-T/2$ and $T/2$. We also omit the subscript n from C, D and A, θ', on the understanding that each of these refers to the same frequency ω.

3.2. Finding the Amplitude

If the initial phase at frequency ω is $-\theta'$ then substituting

$$f(\omega, t) \;=\; A\cos(\omega t - \theta') \tag{3.6}$$

(from Equation 3.2) into Equation 3.5 gives

$$f(\omega)^2 \;=\; \int_t f(\omega, t)\, A\cos(\omega t - \theta')\, dt \tag{3.7}$$

$$\;=\; \int_t A^2 \cos^2(\omega t - \theta')\, dt. \tag{3.8}$$

For any frequency ω, a general result is that (see Section 4.2)

$$\int_{t=-T/2}^{T/2} \cos^2(\omega t)\, dt \;=\; T/2, \tag{3.9}$$

as depicted in Figure 3.2, so Equation 3.8 evaluates to

$$\int_t A^2 \cos^2(\omega t - \theta')\, dt \;=\; A^2 T/2. \tag{3.10}$$

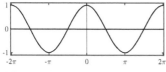

(a) $\cos(\omega t)$ as a function of $\theta = \omega t$

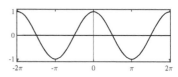

(b) $\cos(\omega t)$ as a function of $\theta = \omega t$

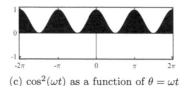

(c) $\cos^2(\omega t)$ as a function of $\theta = \omega t$

Figure 3.2: Representation of Equation 3.9. Both (a) and (b) show $\cos(\omega t)$. (c) The integral of their product $\cos^2(\omega t)$ is the area of the shaded region.

Equating this to Equation 3.7 and solving for A yields

$$A = \frac{2}{T} \int_t f(\omega, t) \cos(\omega t - \theta') \, dt. \qquad (3.11)$$

Thus, if we knew the phase θ' then we could obtain the coefficient A at frequency ω. However, we do not know the phase (yet).

3.3. Finding Sine–Cosine Coefficients

From Equations 3.1 and 3.2 we know that $A \cos(\omega t - \theta') = C \cos(\omega t) + D \sin(\omega t)$, so Equation 3.7 can also be written as

$$f(\omega)^2 = \int_t f(\omega, t) \left(C \cos(\omega t) + D \sin(\omega t) \right) \, dt, \qquad (3.12)$$

which can be split into two integrals

$$f(\omega)^2 = C \int_t f(\omega, t) \cos(\omega t) \, dt + D \int_t f(\omega, t) \sin(\omega t) \, dt. \qquad (3.13)$$

Both integrals can be evaluated using the same approach, so we begin with the first one. Substituting Equation 3.1 into the first integral of

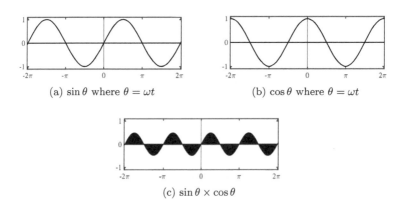

(a) $\sin \theta$ where $\theta = \omega t$ (b) $\cos \theta$ where $\theta = \omega t$

(c) $\sin \theta \times \cos \theta$

Figure 3.3: Representation of Equation 3.20. (a) $\sin(\omega t)$. (b) $\cos(\omega t)$. (c) The height of the curve in the shaded area is $\sin(\omega t) \times \cos(\omega t)$. For every positive height there is a corresponding negative height. Consequently, for every shaded region above zero there is a corresponding shaded region below zero, so the integral of $\sin(\omega t) \times \cos(\omega t)$ is zero.

Equation 3.13 gives

$$C \int_t f(\omega, t) \cos(\omega t)\, dt$$

$$= C \int_t \big[C \cos(\omega t) + D \sin(\omega t) \big] \cos(\omega t)\, dt \tag{3.14}$$

$$= C^2 \int_t \cos(\omega t) \cos(\omega t)\, dt + CD \int_t \sin(\omega t) \cos(\omega t)\, dt. \tag{3.15}$$

In general,

$$\sin \alpha \cos \beta = [\sin(\alpha + \beta) + \sin(\alpha - \beta)]/2. \tag{3.16}$$

If we set $\alpha = \omega t$ and $\beta = \omega t$ then we have

$$\sin(\omega t) \cos(\omega t) = [\sin(\omega t + \omega t) + \sin(\omega t - \omega t)]/2 \tag{3.17}$$
$$= [\sin(2\omega t)]/2. \tag{3.18}$$

Therefore, the second integral in Equation 3.15 can be expressed as

$$\int_t \sin(\omega t) \cos(\omega t)\, dt = \frac{1}{2} \int_t \sin(2\omega t)\, dt. \tag{3.19}$$

By construction, the t-interval $[-T/2, T/2]$ contains an integer number of wavelengths, so the integral on the right-hand side evaluates to zero. Therefore, the second term in Equation 3.15 comes to

$$CD \int_t \cos(\omega t) \sin(\omega t)\, dt = 0, \tag{3.20}$$

as shown in Figure 3.3. Incidentally, this evaluates to zero because $\cos(\omega t)$ and $\sin(\omega t)$ are *orthogonal*, as explained in Chapter 4. Therefore, Equation 3.14 becomes

$$C \int_t f(\omega, t) \cos(\omega t)\, dt = C^2 \int_t \cos^2(\omega t)\, dt. \tag{3.21}$$

Using Equation 3.9, we have

$$C \int_t f(\omega, t) \cos(\omega t)\, dt = C^2 T/2. \tag{3.22}$$

Upon rearranging and dividing through by C we get

$$C = \frac{2}{T} \int_t f(\omega, t) \cos(\omega t)\, dt. \tag{3.23}$$

By analogy, the second integral in Equation 3.13 yields

$$D = \frac{2}{T} \int_t f(\omega, t) \sin(\omega t) \, dt. \tag{3.24}$$

3.4. Finding the Phase

From Equation 2.23, the coefficients C and D can be used to obtain the frequency-specific phase θ'.

We now know how to find the amplitude and phase of a single sinusoid $f(\omega, t)$ at a given frequency ω. Next, we will see how to find the amplitude and phase of a sinusoid when it is part of a complicated signal (i.e. a mixture of sinusoids).

Chapter 4

The Fourier Transform

4.1. Un-mixing Sinusoids

To see how a Fourier transform un-mixes a signal into its component sinusoids, we explore a general strategy for extracting a sinusoid at a single frequency ω from a mixture of sinusoids. Because there is nothing special about the frequency chosen, this general strategy can be applied to all frequencies, which effectively implements the Fourier transform.

4.2. Orthogonal Basis Functions

We wish to represent the function $f(t)$ as a weighted sum of sine and cosine functions. These act as *basis functions*, because they form a basis for representing $f(t)$.

In order to re-create $f(t)$ by adding different basis functions together, it is vital that the frequency information represented by one basis function is not also represented by any other basis function. As we shall see, this condition is guaranteed if the trigonometric basis functions are *mutually orthogonal*.

Orthogonal Cosines

We begin by defining two frequencies,

$$\omega_m = m\omega_1 \quad \text{and} \quad \omega_n = n\omega_1, \tag{4.1}$$

each of which is an integer multiple of the fundamental frequency ω_1. Given cosine functions at frequencies ω_m and ω_n, their *inner product* is defined to be

$$
\begin{aligned}
\cos(\omega_m t) \cdot \cos(\omega_n t) &= \cos(m\omega_1 t) \cdot \cos(n\omega_1 t) \\
&= \int_{t=-T/2}^{T/2} \cos(m\omega_1 t) \cos(n\omega_1 t)\, dt, \tag{4.2}
\end{aligned}
$$

where $T = 2\pi/\omega_1$ (see Equation 2.1). As mentioned earlier, the limits of integration will usually be omitted for brevity. In general, for angles α and β, the product of their cosines is

$$\cos\alpha\cos\beta \quad = \quad [\cos(\alpha + \beta) + \cos(\alpha - \beta)]/2, \tag{4.3}$$

which implies that

$$\cos(m\omega_1 t)\cos(n\omega_1 t) = [\cos((m + n)\omega_1 t) + \cos((m - n)\omega_1 t)]/2. \tag{4.4}$$

Substituting this into Equation 4.2 gives

$$\int_t \cos(m\omega_1 t)\cos(n\omega_1 t)\, dt = \frac{1}{2}\int_t \cos((m + n)\omega_1 t) + \cos((m - n)\omega_1 t)\, dt, \tag{4.5}$$

which can be written as the sum of two integrals,

$$\frac{1}{2}\int_{-T/2}^{T/2} \cos((m + n)\omega_1 t)\, dt + \frac{1}{2}\int_{-T/2}^{T/2} \cos((m - n)\omega_1 t)\, dt. \tag{4.6}$$

We now consider two different scenarios, $m \neq n$ and $m = n$. As a reminder, both m and n are integers, so both ω_m and ω_n are harmonics of the fundamental frequency ω_1.

Case A. If $m \neq n$ then both integrals in Equation 4.6 evaluate to zero, because the integral of a cosine function over a complete number of (i.e.

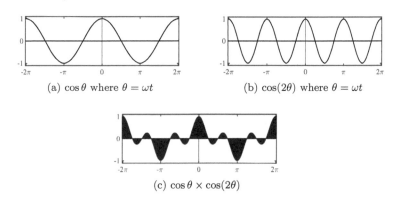

(a) $\cos\theta$ where $\theta = \omega t$

(b) $\cos(2\theta)$ where $\theta = \omega t$

(c) $\cos\theta \times \cos(2\theta)$

Figure 4.1: Illustration of Case A (Equation 4.7). The inner product $\cos(n\omega t)\cdot\cos(m\omega t)$ is zero when $m \neq n$. (a) $\cos(n\omega t)$; (b) $\cos(m\omega t)$ (here $n = 1$ and $m = 2$); (c) $\cos(n\omega t)\times\cos(m\omega t)$. For every positive height there is a corresponding negative height. Consequently, for every shaded region above zero there is a corresponding shaded region below zero, so the integral of $\cos(n\omega t)\times\cos(m\omega t)$ is zero.

$m + n$ or $m - n$) periods is zero. Thus, we have established that cosine functions with different harmonics (i.e. $m \neq n$) are orthogonal:

$$\int_{-T/2}^{T/2} \cos(m\omega_1 t) \cos(n\omega_1 t)\, dt \quad = \quad 0, \qquad (4.7)$$

as illustrated in Figure 4.1.

Case B. In contrast, if $m = n$ then the first integral in Equation 4.6 evaluates to zero — again because the integral of a cosine function over a complete number of (i.e. $2n$) periods is zero. But in the second integral in Equation 4.6, $\cos((m - n)\omega_1 t) = \cos 0 = 1$, so the integral evaluates to T and hence Equation 4.6 evaluates to $T/2$. Therefore, when $m = n$, Equation 4.5 becomes

$$\int_{-T/2}^{T/2} \cos^2(m\omega_1 t)\, dt \quad = \quad T/2, \qquad (4.8)$$

which will prove useful shortly. The equality in Equation 4.8 is illustrated in Figure 3.2. In summary, the inner product

$$\cos(\omega_m t) \cdot \cos(\omega_n t) \quad = \quad 0 \quad \text{unless } m = n. \qquad (4.9)$$

Thus, two cosine functions with frequencies that are different multiples of the same fundamental frequency are orthogonal.

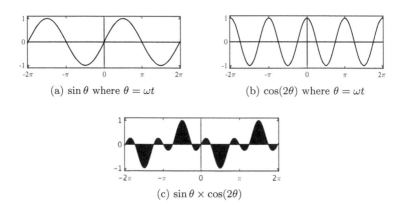

(a) $\sin\theta$ where $\theta = \omega t$

(b) $\cos(2\theta)$ where $\theta = \omega t$

(c) $\sin\theta \times \cos(2\theta)$

Figure 4.2: A sine–cosine inner product $\sin(n\omega t) \cdot \cos(m\omega t)$ (Equation 4.10) with $m \neq n$ equals zero. (a) $\sin(n\omega t)$; (b) $\cos(m\omega t)$ (here $n = 1$ and $m = 2$); (c) $\sin(n\omega t) \times \cos m\omega t$. For every positive height there is a corresponding negative height, which cancel, so the integral of $\sin(n\omega t) \times \cos(m\omega t)$ is zero.

Sines and Cosines are Orthogonal

We use a similar procedure to test whether sines and cosines are orthogonal. The inner product of a sine and a cosine is

$$\sin(\omega_n t) \cdot \cos(\omega_m t) \quad = \quad \int_{-T/2}^{T/2} \sin(n\omega_1 t) \cos(m\omega_1 t) \, dt. \qquad (4.10)$$

In general,

$$\sin \alpha \cos \beta \quad = \quad [\sin(\alpha + \beta) + \sin(\alpha - \beta)]/2. \qquad (4.11)$$

If we set $\alpha = n\omega_1 t$ and $\beta = m\omega_1 t$ then we have

$$\sin(n\omega_1 t) \cos(m\omega_1 t) \quad = \quad \left[\sin\big((m + n)\omega_1 t\big) + \sin\big((m - n)\omega_1 t\big)\right]/2, \qquad (4.12)$$

which allows us to express Equation 4.10 as two integrals,

$$\frac{1}{2} \int_{-T/2}^{T/2} \sin\big((m + n)\omega_1 t\big) \, dt + \frac{1}{2} \int_{-T/2}^{T/2} \sin\big((m - n)\omega_1 t\big) \, dt. \qquad (4.13)$$

By construction, both integrals extend over an integer number of wavelengths and so evaluate to zero; therefore

$$\sin(\omega_m t) \cdot \cos(\omega_n t) \quad = \quad 0. \qquad (4.14)$$

The integral of Equation 4.10 is illustrated for $m \neq n$ in Figure 4.2 and for $m = n$ in Figure 3.3, and is zero in both cases.

4.3. Finding Fourier Coefficients

How can we find C_n and D_n? We begin by finding C_n for a particular frequency ω_n. The basic strategy consists of finding the projection of the function $f(t)$ onto the sinusoid with frequency ω_n. This is achieved by multiplying $f(t)$ in Equation 2.18 (repeated here),

$$f(t) \quad = \quad C_0/2 + \sum_{m=1}^{\infty} C_m \cos(\omega_m t) + D_m \sin(\omega_m t), \qquad (4.15)$$

by $\cos(\omega_n t)$ and integrating over t between $-T/2$ and $T/2$:

$$\int_t f(t) \cos(\omega_n t) \, dt = \int_t \left[\sum_{m=1}^{\infty} C_m \cos(\omega_m t) + D_m \sin(\omega_m t)\right] \cos(\omega_n t) \, dt, \qquad (4.16)$$

where we ignore C_0 for now and recall that

$$\sin(\omega_n t)\cos(\omega_m t) \;=\; \sin(n\omega_1 t)\cos(m\omega_1 t). \qquad (4.17)$$

Moving $\cos(\omega_n t)$ inside the square brackets yields two integrals,

$$\int_t \left[\sum_{m=1}^{\infty} C_m \cos(\omega_m t)\cos(\omega_n t)\right] dt + \int_t \left[\sum_{m=1}^{\infty} D_m \sin(\omega_m t)\cos(\omega_n t)\right] dt.$$
$$(4.18)$$

For convenience, we swap the summations and integrals (this can be done, though we omit the justification here for brevity) to get

$$\sum_{m=1}^{\infty} C_m \int_t \cos(\omega_m t)\cos(\omega_n t)\, dt + \sum_{m=1}^{\infty} D_m \int_t \sin(\omega_m t)\cos(\omega_n t)\, dt.$$
$$(4.19)$$

Using Equation 4.14, each integral in the second summation evaluates to zero. Therefore, Equation 4.16 becomes

$$\int_t f(t)\cos(\omega_n t)\, dt \;=\; \sum_{m=1}^{\infty} C_m \int_t \cos(\omega_m t)\cos(\omega_n t)\, dt. \qquad (4.20)$$

As above, we consider the different cases of $m \neq n$ and $m = n$.

Case 1. For each $m \neq n$, the corresponding integral in Equation 4.20 equals zero (see Equation 4.7). Thus, only one term from the summation survives, the one with $m = n$.

Case 2. When $m = n$ we have

$$\int_t f(t)\cos(\omega_n t)\, dt \;=\; C_n \int_t \cos^2(\omega_n t)\, dt, \qquad (4.21)$$

where (from Equation 4.8)

$$\int_t \cos^2(\omega_n t)\, dt \;=\; \frac{T}{2}. \qquad (4.22)$$

Therefore, Equation 4.21 becomes

$$\int_t f(t)\cos(\omega_n t)\, dt \;=\; C_n \frac{T}{2}. \qquad (4.23)$$

Rearranging this yields

$$C_n \;=\; \frac{2}{T}\int_t f(t)\cos(\omega_n t)\, dt. \qquad (4.24)$$

By analogy,

$$D_n = \frac{2}{T} \int_t f(t) \sin(\omega_n t)\, dt. \tag{4.25}$$

The mean value of $f(t)$ is represented by the coefficients of the frequency corresponding to $n = 0$, which is $\omega_0 = 0$ (see Section 2.3). In Equation 4.25, $\sin(\omega_0 t) = \sin 0 = 0$, so we have $D_0 = 0$. In contrast, in Equation 4.24 we have $\cos(\omega_0 t) = \cos 0 = 1$, and therefore

$$C_0 = \frac{2}{T} \int_t f(t)\, dt, \tag{4.26}$$

which is twice the mean value of $f(t)$. This is why C_0 is divided by 2 in Equation 4.15.

More generally, the orthogonality properties described above mean that multiplying a superposition of sinusoids $f(t)$ by a single sinusoid $f(\omega, t) = \cos(\omega t - \theta'_\omega)$ has the effect of 'picking out' the component of $f(t)$ that has the same frequency and phase as $f(\omega, t)$. For example, suppose we know the value of θ' at frequency ω_n, which we write as θ'_n. If we were to replace $\cos(\omega_n t)$ with $\cos(\omega_n t - \theta'_n)$ in Equation 4.16 then

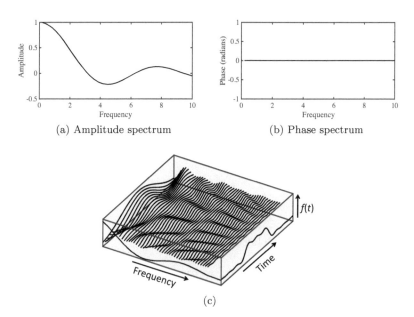

(a) Amplitude spectrum   (b) Phase spectrum

(c)

Figure 4.3: (a) Amplitude spectrum, also shown on the front panel of (c). (b) Phase spectrum, in which the phase at every frequency is a constant $\theta'_n = 0$. (c) How sinusoids with the amplitude and phase spectra in (a) and (b) combine to form the function $f(t)$ on the right end panel.

the result would be

$$A_n = \frac{2}{T} \int_t f(t) \cos(\omega_n t - \theta'_n) \, dt, \qquad (4.27)$$

where C_n in Equation 4.24 is replaced by A_n, and $A_0 = C_0$ from Equation 4.26.

4.4. Amplitude and Phase Spectra

Irrespective of whether we choose to represent the Fourier transform of a signal as a set of amplitudes and phases $\{A, \theta'\}$ or as a set of pairs of coefficients $\{C, D\}$, every Fourier component must be specified by two distinct numbers. In other words, it is not sufficient to know 'how much' (i.e. the amplitude A) of each frequency component is present in a signal; we also need to know the phase of each component. For this reason, plotting the graph of a Fourier transform can be misleading, because it usually displays information only about the amplitude spectrum. Similarly, the set of phases of a Fourier transform is the *phase spectrum*.

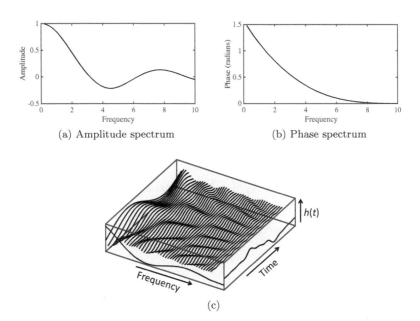

(a) Amplitude spectrum (b) Phase spectrum

(c)

Figure 4.4: (a) Amplitude spectrum, which is also shown on the front panel of (c) and is the same as in Figure 4.3. (b) Phase spectrum, in which the phase decreases with frequency. (c) How sinusoids with the amplitude and phase spectra in (a) and (b) combine to form the function $h(t)$ on the right end panel.

As an example, the amplitude and phase spectra of a function $f(t)$ are shown in Figure 4.3a and b, respectively, and the function $f(t)$ itself is plotted on the right end panel of Figure 4.3c. By design, the function $h(t)$ in Figure 4.4 has the same amplitude spectrum as $f(t)$, defined by $A_n = \sin(\omega_n t)/(\omega_n t)$ in both cases (also see Equation 8.34). But whereas the phase of $f(t)$ is zero at all frequencies (as shown in Figure 4.3b), the phase spectrum of $h(t)$ has the profile shown in Figure 4.4b. As a result the function $h(t)$ (shown on the right end panel of Figure 4.4d) looks like a distorted version of the function $f(t)$. For analogous examples with images, see Figures 7.7 and 8.1.

4.5. Summary

The inverse Fourier transform of a function $f(t)$ is (from Equation 2.18)

$$f(t) \;=\; C_0/2 + \sum_{n=1}^{\infty} C_n \cos(\omega_n t) + D_n \sin(\omega_n t), \qquad (4.28)$$

where the term $C_0/2$ is added to accommodate functions with nonzero means (see Section 4.3). The Fourier transform of $f(t)$ consists of

$$C_n \;=\; \frac{2}{T} \int_{t=-T/2}^{T/2} f(t) \cos(\omega_n t)\, dt, \qquad (4.29)$$

$$D_n \;=\; \frac{2}{T} \int_{t=-T/2}^{T/2} f(t) \sin(\omega_n t)\, dt. \qquad (4.30)$$

Alternatively and equivalently, the inverse Fourier transform is (from Equation 2.19)

$$f(t) \;=\; A_0/2 + \sum_{n=1}^{\infty} A_n \cos(\omega_n t - \theta'_n), \qquad (4.31)$$

where the term $A_0/2$ accommodates functions with nonzero means.

The amplitude spectrum of the Fourier transform of $f(t)$ is (from Equation 4.27)

$$A_n \;=\; \frac{2}{T} \int_{t=-T/2}^{T/2} f(t) \cos(\omega_n t - \theta'_n)\, dt. \qquad (4.32)$$

Chapter 5

Visualising Complex Waves

5.1. Complex Numbers

The preceding equations can be expressed more succinctly using complex numbers. A complex number consists of two parts, a *real number* and an *imaginary number*. The real part is a conventional number, like 3 or 4.2 or π. The imaginary part is a multiple of the *unit imaginary number* i, where i is the square root of minus one, so that

$$i^2 = -1. \tag{5.1}$$

A complex number $\hat{z}$ and its *complex conjugate* $\hat{z}^*$ are

$$
\begin{aligned}
\hat{z} &= x + iy, \\
\hat{z}^* &= x - iy,
\end{aligned}
\tag{5.2}
$$

where $x = \text{Re}(\hat{z})$ is real and $iy = i \times \text{Im}(\hat{z})$ is a *pure imaginary number*, which is equal to y lots of the unit imaginary number i. Here we will write all complex numbers with a hat symbol.

Addition

$$
\begin{aligned}
\hat{z}_1 + \hat{z}_2 &= (x_1 + iy_1) + (x_2 + iy_2) \\
&= (x_1 + x_2) + i(y_1 + y_2).
\end{aligned}
\tag{5.3}
$$

Subtraction

$$
\begin{aligned}
\hat{z}_1 - \hat{z}_2 &= (x_1 + iy_1) - (x_2 + iy_2) \\
&= (x_1 - x_2) + i(y_1 - y_2).
\end{aligned}
\tag{5.4}
$$

Whereas addition and subtraction are straightforward, multiplication and division are not.

Multiplication

$$\hat{z}_1\hat{z}_2 = (x_1 + iy_1) \times (x_2 + iy_2)$$
$$= (x_1x_2 - y_1y_2) + i(x_1y_2 + y_1x_2). \tag{5.5}$$

Division

$$\hat{z}_1\hat{z}_2 = \frac{x_1 + iy_1}{x_2 + iy_2} = \frac{x_1 + iy_1}{x_2 + iy_2} \times \frac{x_2 - iy_2}{x_2 - iy_2}$$
$$= \frac{x_1x_2 + y_1y_2}{x_2^2 + y_2^2} + i\frac{x_2y_1 - x_1y_2}{x_2^2 + y_2^2}. \tag{5.6}$$

Fortunately, Euler's theorem (introduced below) allows multiplication and division to be simplified.

5.2. The Complex Plane

A complex number can be represented as a point on the *complex plane*, also known as an *Argand diagram* (Figure 5.1). The real part is represented on the horizontal axis, and the imaginary part is represented on the vertical axis, so that $\hat{z}$ is represented by the *Cartesian coordinates* (x, y). It can also be represented in *polar coordinates* (A, θ), where A is the length of the vector from the origin of the Argand diagram to $\hat{z}$ and θ is the angle this vector makes with the positive x-axis. The squared length of the vector is

$$A^2 = \hat{z}\hat{z}^*$$
$$= (x + iy)(x - iy)$$
$$= x^2 - i^2y^2 + ixy - ixy, \tag{5.7}$$

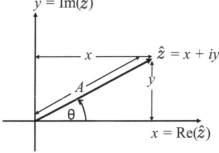

Figure 5.1: Argand diagram representation of a complex number $\hat{z}$.

where $-i^2 = 1$, so that

$$A^2 \;=\; x^2 + y^2. \tag{5.8}$$

Therefore, the length, *modulus* or *magnitude* of the complex number $\hat{z}$ is

$$\begin{aligned} A \;&=\; |\hat{z}| \;=\; \sqrt{\hat{z}\hat{z}^*} \\ &=\; \sqrt{x^2 + y^2}. \end{aligned} \tag{5.9}$$

The angle θ is called the *phase*, and the ratio of the imaginary part to the real part of $\hat{z}$ is $y/x = \tan\theta$. From Figure 5.1, the real and imaginary parts of $\hat{z}$ can be expressed as

$$\begin{aligned} x \;&=\; A\cos\theta, \\ y \;&=\; A\sin\theta. \end{aligned} \tag{5.10}$$

Substituting these in Equation 5.2 yields the complex sinusoidal function

$$\hat{z} \;=\; A(\cos\theta + i\sin\theta). \tag{5.11}$$

5.3. Euler's Theorem

We introduce *Euler's theorem*, which states that

$$e^{i\theta} \;=\; \cos\theta + i\sin\theta. \tag{5.12}$$

It is worth noting that $e^{-i\theta}$ is the complex conjugate of $e^{i\theta}$. Specifically, $e^{-i\theta} = e^{i(-\theta)} = \cos(-\theta) + i\sin(-\theta)$, and because $\cos(-\theta) = \cos\theta$ (cosine is a even function) and $\sin(-\theta) = -\sin\theta$ (sine is an odd function),

$$e^{-i\theta} \;=\; \cos\theta - i\sin\theta. \tag{5.13}$$

Therefore, if $\hat{z} = Ae^{i\theta}$ then $Ae^{-i\theta}$ is the complex conjugate of $\hat{z}$:

$$\begin{aligned} \hat{z} \;&=\; Ae^{i\theta}, \tag{5.14} \\ \hat{z}^* \;&=\; Ae^{-i\theta}. \tag{5.15} \end{aligned}$$

It will prove useful to note that when a complex number and its conjugate are added together, the imaginary parts cancel:

$$\begin{aligned} \hat{z} + \hat{z}^* \;&=\; Ae^{i\theta} + Ae^{-i\theta} \\ &=\; A(\cos\theta + i\sin\theta) + A(\cos\theta - i\sin\theta) \\ &=\; 2A\cos\theta. \end{aligned} \tag{5.16}$$

Also, when a complex number is subtracted from its conjugate, the real parts cancel:

$$
\begin{aligned}
\hat{z} - \hat{z}^* &= Ae^{i\theta} - Ae^{-i\theta} \\
&= A(\cos\theta + i\sin\theta) - A(\cos\theta - i\sin\theta) \\
&= 2Ai\sin\theta.
\end{aligned}
\tag{5.17}
$$

Expressing a complex variable in terms of exponents greatly simplifies the multiplication and division of complex numbers, which would otherwise involve much tedious manipulation of sines and cosines.

Multiplication

Given two complex numbers

$$
\hat{z}_1 = A_1 e^{i\theta_1} \quad \text{and} \quad \hat{z}_2 = A_2 e^{i\theta_2},
\tag{5.18}
$$

multiplication by $\hat{z}_2$ rotates $\hat{z}_1$ through an angle of *plus* θ_2:

$$
\hat{z}_1 \hat{z}_2 = A_1 e^{i\theta_1} \times A_2 e^{i\theta_2} = A_1 A_2 e^{i(\theta_1 + \theta_2)},
\tag{5.19}
$$

where $+\theta_2$ is a rotation in the *anti-clockwise direction*.

Division

Dividing $\hat{z}_1$ by $\hat{z}_2$ effectively rotates $\hat{z}_1$ through an angle of *minus* θ_2:

$$
\frac{\hat{z}_1}{\hat{z}_2} = \frac{A_1 e^{i\theta_1}}{A_2 e^{i\theta_2}} = \frac{A_1}{A_2} e^{i(\theta_1 - \theta_2)},
\tag{5.20}
$$

where $-\theta_2$ is a rotation in the *clockwise direction* on the complex plane.

5.4. Visualising Complex Waves

The Anti-clockwise Travelling Clock

As in Chapter 1, we can express the angle θ in terms of the angular frequency ω and time t as $\theta = \omega t$. We define the complex function

$$
\begin{aligned}
\hat{f}_\omega(t) &= \cos(\omega t) + i\sin(\omega t) \tag{5.21} \\
&= e^{i\omega t}. \tag{5.22}
\end{aligned}
$$

This can be visualised as a helix generated by a clock where the clock hand rotates anti-clockwise as the clock moves through space, as shown in Figure 5.2.

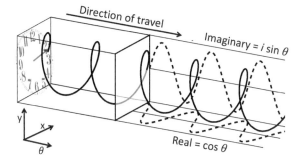

Figure 5.2: The complex function $\hat{f}(t) = e^{i\omega t}$ represented as an anti-clockwise helix, generated by a clock hand which rotates at ω rad/s (Equation 5.22). In the box on the left, only $\hat{f}(t)$ is shown. On the right, the real part of $\hat{f}(t)$ is the cosine function (dashed) on the ground plane, and the imaginary part is the sine function (dashed) on the vertical plane.

We assume that the clock moves at constant speed but the clock hand can rotate at any rate. This means that the 'tightness' of the helix depends only on how quickly the clock hand rotates; the faster it rotates, the tighter the helix. The rate at which the clock hand rotates is its angular frequency.

If the clock hand starts at 3 o'clock, and if the helix casts a shadow on the ground, then one helix loop corresponds to one cycle of a cosine wave on the ground, where the distance between consecutive peaks of the wave is the wavelength λ. Similarly, if the helix casts a shadow on a vertical plane then the result is a sine wave with the same wavelength λ.

The phase of the complex wave defined in Equation 5.22 is $\theta = \omega t$, and its amplitude is

$$|\hat{f}_\omega(t)| = \sqrt{\hat{f}_\omega(t)\,\hat{f}_\omega^*(t)} = \sqrt{e^{i\omega t}\,e^{-i\omega t}} = 1. \qquad (5.23)$$

The Clockwise Travelling Clock

If the clock hand starts at the number 3 on the clock face and rotates clockwise then the helix defined by the clock hand also turns in a clockwise direction, as shown in Figure 5.3. In effect, what was t in Equation 5.21 becomes $-t$, so we have

$$\hat{f}_\omega(-t) = e^{i\omega(-t)} \qquad (5.24)$$
$$= \cos(\omega \times (-t)) + i\sin(\omega \times (-t)) \qquad (5.25)$$
$$= \cos(\omega t) - i\sin(\omega t). \qquad (5.26)$$

Note that $e^{i\omega(-t)} = e^{-i\omega t} = e^{i(-\omega)t}$, so as a function of t (rather than of $-t$) we will write this complex wave as $\hat{f}_{-\omega}(t)$. The phase of this

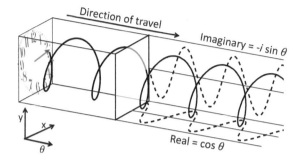

Figure 5.3: The complex function $\hat{f}(t) = e^{-i\omega t}$ represented as a clockwise helix which rotates at the rate of ω rad/s (Equation 5.26).

wave is $\theta = -\omega t$, and its amplitude is

$$|\hat{f}_{-\omega}(t)| = \sqrt{e^{-i\omega t}\, e^{i\omega t}} = 1. \tag{5.27}$$

In summary, the anti-clockwise and clockwise complex waves $\hat{f}_\omega(t)$ and $\hat{f}_{-\omega}(t)$ both have an amplitude of 1. Because they have the same frequency, their real parts are the same while their imaginary parts have the same magnitude but opposite signs. Therefore, the imaginary parts cancel when the clockwise and anti-clockwise waves are added together,

$$\hat{f}_\omega(t) + \hat{f}_{-\omega}(t) = \big(\cos(\omega t) + i\sin(\omega t)\big) + \big(\cos(\omega t) - i\sin(\omega t)\big) \tag{5.28}$$
$$= 2\cos(\omega t), \tag{5.29}$$

as shown in Figure 5.4.

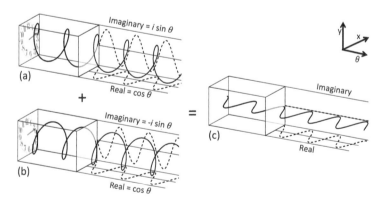

Figure 5.4: If an anti-clockwise complex wave (a) is added to a clockwise complex wave (b) then the imaginary parts in the vertical plane cancel, yielding a real cosine wave (c) in the horizontal plane with twice the amplitude of each of the waves in (a) and (b) (see Equation 5.29).

In Equations 5.24–5.26 we chose to associate the negative sign with t, but clockwise rotation is often interpreted as an angular frequency of $-\omega$ (the product of angular frequency and time is negative in both cases). Consequently, what we call clockwise helices here are often referred to as *negative* frequencies, which is (to say the least) counter-intuitive. In fact, so-called positive and negative frequencies can also be interpreted as complex helices that rotate in opposite directions.

5.5. Setting the Phase and Amplitude

The amplitude and initial phase of an anti-clockwise sinusoidal wave with frequency ω can be set by a complex constant $\hat{A}_\omega$. We now define

$$\hat{f}_\omega(t) \;\; = \;\; \hat{A}_\omega \, e^{i\omega t}, \tag{5.30}$$

where (as will be proved in Section 6.1)

$$\hat{A}_\omega \;\; = \;\; (C_\omega - iD_\omega)/2. \tag{5.31}$$

Expanding both terms in Equation 5.30, we get

$$\hat{f}_\omega(t) \;\; = \;\; [C_\omega \, e^{i\omega t} - iD_\omega \, e^{i\omega t}]/2 \tag{5.32}$$

$$= \;\; \left[C_\omega \big(\cos(\omega t) + i\sin(\omega t)\big) - iD_\omega \big(\cos(\omega t) + i\sin(\omega t)\big) \right]/2 \tag{5.33}$$

$$= \;\; \left[C_\omega \cos(\omega t) + iC_\omega \sin(\omega t) - iD_\omega \cos(\omega t) - i^2 D_\omega \sin(\omega t) \right]/2, \tag{5.34}$$

where $-i^2 = 1$, so in terms of real and imaginary component functions,

$$\hat{f}_\omega(t) = [C_\omega \cos(\omega t) + D_\omega \sin(\omega t)]/2 + i[C_\omega \sin(\omega t) - D_\omega \cos(\omega t)]/2.$$

Similarly, the amplitude and initial phase of a clockwise complex sinusoidal wave can be set by a complex constant $\hat{A}_{-\omega}$:

$$\hat{f}_{-\omega}(t) \;\; = \;\; \hat{A}_{-\omega} \, e^{-i\omega t}, \tag{5.35}$$

where (as will be proved in Section 6.1)

$$\hat{A}_{-\omega} \;\; = \;\; (C_\omega + iD_\omega)/2. \tag{5.36}$$

Therefore Equation 5.35 can be written as

$$\hat{f}_{-\omega}(t) = [C_\omega \cos(\omega t) + D_\omega \sin(\omega t)]/2 - i[C_\omega \sin(\omega t) - D_\omega \cos(\omega t)]/2.$$

Now a real frequency component is obtained as the sum of two complex components (where the imaginary parts cancel),

$$f_\omega(t) = \hat{f}_\omega(t) + \hat{f}_{-\omega}(t) \tag{5.37}$$
$$= \hat{A}_\omega\, e^{i\omega t} + \hat{A}_{-\omega}\, e^{-i\omega t} \tag{5.38}$$
$$= C_\omega \cos(\omega t) + D_\omega \sin(\omega t). \tag{5.39}$$

Notice this is the sum of a clockwise and an anti-clockwise complex function (as in Equations 5.16 and 5.29), which will prove useful later.

The amplitude of $\hat{f}_\omega(t)$ is given by its modulus,

$$|\hat{f}_\omega(t)| = |\hat{A}_\omega\, e^{i\omega t}|. \tag{5.40}$$

The modulus of a product of complex numbers is equal to the product of their moduli,

$$|\hat{f}_\omega(t)| = |\hat{A}_\omega|\,|e^{i\omega t}|, \tag{5.41}$$

which can be obtained from the squared moduli,

$$|\hat{f}_\omega(t)|^2 = |\hat{A}_\omega|^2\,|e^{i\omega t}|^2. \tag{5.42}$$

We know from Equation 5.23 that $|e^{i\omega t}|^2 = 1$, so Equation 5.42 becomes

$$|\hat{f}_\omega(t)|^2 = |\hat{A}_\omega|^2 = \hat{A}_\omega \hat{A}_\omega^*, \tag{5.43}$$

where $\hat{A}_\omega = (C_\omega - iD_\omega)/2$ and $\hat{A}_\omega^* = (C_\omega + iD_\omega)/2$, so that

$$|\hat{A}_\omega|^2 = (C_\omega - iD_\omega)/2 \times (C_\omega + iD_\omega)/2 \tag{5.44}$$
$$= (C_\omega^2 + D_\omega^2)/4. \tag{5.45}$$

Therefore the amplitude of the complex function $\hat{f}_\omega(t)$ is

$$|\hat{f}_\omega(t)| = |\hat{A}_\omega| = \sqrt{C_\omega^2 + D_\omega^2}\,/2. \tag{5.46}$$

By symmetry, the amplitude of $\hat{f}_{-\omega}(t)$ is

$$|\hat{f}_{-\omega}(t)| = |\hat{A}_{-\omega}| = \sqrt{C_\omega^2 + D_\omega^2}\,/2. \tag{5.47}$$

It will prove useful to note that the amplitude of $f_\omega(t)$ at frequency ω is the sum of the complex function amplitudes at ω and at $-\omega$,

$$A_\omega = |\hat{A}_\omega| + |\hat{A}_{-\omega}| = \sqrt{C_\omega^2 + D_\omega^2}. \tag{5.48}$$

The initial phases of the real and imaginary parts of the sinusoid are

$$\theta_\omega^{\text{Real}} = \arctan(D_\omega/C_\omega), \tag{5.49}$$
$$\theta_\omega^{\text{Imag}} = \arctan(-C_\omega/D_\omega). \tag{5.50}$$

Chapter 6

The Complex Fourier Transform

6.1. Mixing Complex Waves

Before we can understand how to un-mix complex waves, we need to know how complex waves get mixed together. Accordingly, consider a periodic function $f_T(t)$ that repeats every T seconds. The lowest nonzero angular frequency (i.e. the fundamental frequency) is

$$\omega_1 \quad = \quad 2\pi/T \text{ rad/s}. \tag{6.1}$$

If we omit the 1 subscript from ω_1, and omit the T subscript from $f_{T(t)}$, (to minimise notational clutter), and replace ω_n with $n\omega$ then Equation 2.18 becomes

$$f(t) \quad = \quad C_0/2 + \sum_{n=1}^{\infty} C_n \cos(n\omega t) + D_n \sin(n\omega t). \tag{6.2}$$

Our objective is to express $f(t)$ in terms of complex numbers as

$$f(t) \quad = \quad \sum_{n=-\infty}^{\infty} \hat{A}_n e^{in\omega t} \tag{6.3}$$

and map the coefficients C_n and D_n to the complex coefficients $\hat{A}_n$. Replacing the ω subscript in Equation 5.39 with n (simplified from ω_n),

$$f_n(t) \quad = \quad C_n \cos(n\omega t) + D_n \sin(n\omega t), \tag{6.4}$$

so that Equation 6.2 can be written as

$$f(t) \quad = \quad C_0/2 + \sum_{n=1}^{\infty} f_n(t). \tag{6.5}$$

Splitting Infinity. Equation 6.3 can be split into two summations, one spanning positive values of n and the other spanning negative values,

$$f(t) \quad = \quad \hat{A}_0 e^{i\omega_0 t} + \sum_{n=-\infty}^{-1} \hat{A}_n e^{in\omega t} + \sum_{n=1}^{\infty} \hat{A}_n e^{in\omega t}. \qquad (6.6)$$

Since $\omega_0 = 0$, we have $e^{i\omega_0 t} = 1$, so comparing the constant terms in Equations 6.2 and 6.6 gives $C_0/2 = \hat{A}_0$. Therefore the amplitude $\hat{A}_0$ at $\omega_0 = 0$ is a real number $A_0/2$ where $A_0 = C_0$ in Equation 6.2.

We rewrite the first summation in Equation 6.6 by changing the index values so that they span the range from $n = 1$ to ∞ and, accordingly, replacing n with $-n$ in each summed term:

$$f(t) \quad = \quad A_0/2 + \sum_{n=1}^{\infty} \hat{A}_{-n} e^{-in\omega t} + \sum_{n=1}^{\infty} \hat{A}_n e^{in\omega t}. \qquad (6.7)$$

Because both summations employ values of n from 1 to ∞, we have

$$f(t) \quad = \quad A_0/2 + \sum_{n=1}^{\infty} (\hat{A}_n e^{in\omega t} + \hat{A}_{-n} e^{-in\omega t}). \qquad (6.8)$$

Comparison with Equation 6.2 implies that

$$\hat{A}_n e^{in\omega t} + \hat{A}_{-n} e^{-in\omega t} \quad = \quad C_\omega \cos(\omega t) + D_\omega \sin(\omega t), \qquad (6.9)$$

and therefore Equation 6.4 can be written as

$$f_n(t) \quad = \quad \hat{A}_n e^{in\omega t} + \hat{A}_{-n} e^{-in\omega t}, \qquad (6.10)$$

which is the sum of clockwise and anti-clockwise complex functions at one frequency (see Equation 5.38 and Figure 5.4). Using Equation 5.48 with a change of subscript, the amplitude A_n of $f_n(t)$ is the sum of complex function amplitudes, $A_n = |\hat{A}_n| + |\hat{A}_{-n}|$.

From Real to Complex Coefficients. From Equations 5.12 and 5.13,

$$e^{in\omega t} \quad = \quad \cos(n\omega t) + i \sin(n\omega t), \qquad (6.11)$$
$$e^{-in\omega t} \quad = \quad \cos(n\omega t) - i \sin(n\omega t). \qquad (6.12)$$

Substituting these into Equation 6.10,

$$f_n(t) \quad = \quad \hat{A}_n [\cos(n\omega t) + i \sin(n\omega t)] + \hat{A}_{-n} [\cos(n\omega t) - i \sin(n\omega t)]$$

$$= \quad (\hat{A}_n + \hat{A}_{-n}) \cos(n\omega t) + i(\hat{A}_n - \hat{A}_{-n}) \sin(n\omega t). \qquad (6.13)$$

Comparison with Equation 6.4 implies that

$$C_n = (\hat{A}_n + \hat{A}_{-n}), \tag{6.14}$$
$$D_n = i(\hat{A}_n - \hat{A}_{-n}). \tag{6.15}$$

Finally, solving for $\hat{A}_n$ and $\hat{A}_{-n}$ yields

$$\hat{A}_n = (C_n - iD_n)/2, \tag{6.16}$$
$$\hat{A}_{-n} = (C_n + iD_n)/2. \tag{6.17}$$

6.2. The Complex Fourier Transform

Here, we derive an expression that allows the complex coefficients $\hat{A}_n$ to be calculated from the function $f(t)$. From Section 4.3 we know that

$$C_n = \frac{2}{T} \int_{-T/2}^{T/2} f(t) \cos(\omega_n t)\, dt, \tag{6.18}$$

$$D_n = \frac{2}{T} \int_{-T/2}^{T/2} f(t) \sin(\omega_n t)\, dt, \tag{6.19}$$

and from Equation 6.16 we know that

$$\hat{A}_n = \frac{1}{2}(C_n - iD_n). \tag{6.20}$$

Therefore, substituting Equations 6.18 and 6.19 into Equation 6.20,

$$\hat{A}_n = \frac{1}{2}\left(\frac{2}{T} \int_{-T/2}^{T/2} f(t) \cos(\omega_n t)\, dt - i\frac{2}{T} \int_{-T/2}^{T/2} f(t) \sin(\omega_n t)\, dt\right)$$

$$= \frac{1}{T} \int_{-T/2}^{T/2} f(t) \left[\cos(\omega_n t) - i\sin(\omega_n t)\right] dt. \tag{6.21}$$

Using Equation 6.12 to write the expression in square brackets as $e^{i\omega_n t}$ and then dropping the n subscript, we obtain the *complex Fourier transform* of $f(t)$,

$$\hat{A}_\omega = \frac{1}{T} \int_{-T/2}^{T/2} f(t)\, e^{-i\omega t}\, dt, \tag{6.22}$$

where $\hat{A}_n$ has been rewritten as $\hat{A}_\omega$. In terms of frequency, this is

$$\hat{A}_\nu = \frac{1}{T} \int_{-T/2}^{T/2} f(t)\, e^{-i2\pi\nu t}\, dt. \tag{6.23}$$

41

6.3. Fourier Pairs

The Fourier transform is conventionally written in terms of pairs of integrals, called *Fourier pairs*, which can take several equivalent forms.

The Fourier Pair in Terms of ω. Equation 6.3 can be written as

$$f(t) \;=\; \sum_{n=-\infty}^{\infty} \hat{A}_\omega \, e^{i\omega_n t} \, \Delta n, \tag{6.24}$$

where $\hat{A}_n$ has been changed to $\hat{A}_\omega$ as in Equation 6.22, and where $\Delta n = 1$ is made explicit here because we intend to convert Equation 6.24 into an integral. We know that $n = \omega_n/\omega_1$, so

$$\Delta n \;=\; \Delta\omega/\omega_1, \tag{6.25}$$

where $\omega_1 = 2\pi/T$ is the fundamental frequency and $\Delta\omega$ is an increment in frequency. It follows that

$$\begin{aligned} \Delta\omega \;&=\; \omega_1 \Delta n \\ &=\; 2\pi \Delta n/T \end{aligned} \tag{6.26}$$

and so $\Delta n = T\Delta\omega/(2\pi)$. Substituting this into Equation 6.24 gives

$$f(t) \;=\; \frac{T}{2\pi} \sum_{\omega=-\infty}^{\infty} \hat{A}_\omega \, e^{i\omega t} \, \Delta\omega. \tag{6.27}$$

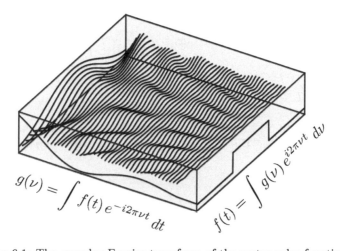

Figure 6.1: The complex Fourier transform of the rectangular function $f(t)$ on the right end panel, expressed in terms of the temporal frequency ν. Here $f(t)$ is the sum of an infinite number of Fourier components.

From Equation 6.26, notice that as $T \to \infty$, so $\Delta\omega \to 0$, and therefore every frequency will occur in the summation of Equation 6.27.

Now, to obtain the two defining relationships of the Fourier transform, we define $g(\omega) = T\hat{A}_\omega$, so that

$$\hat{A}_\omega = g(\omega)/T. \tag{6.28}$$

Substituting this into Equation 6.27 and letting $T \to \infty$ so that $\Delta\omega \to 0$ yields the *Fourier integral*

$$f(t) = \frac{1}{2\pi} \int_{\omega=-\infty}^{\infty} g(\omega)\, e^{i\omega t}\, d\omega. \tag{6.29}$$

Substituting Equation 6.28 into Equation 6.22 gives

$$g(\omega) = \int_{t=-\infty}^{\infty} f(t)\, e^{-i\omega t}\, dt. \tag{6.30}$$

Equations 6.29 and 6.30 constitute the Fourier pair in terms of ω. Because the product of the angular frequency ω and time t is dimensionless, these are *conjugate variables* (see Section 7.3).

The Fourier Pair in Terms of ν. Equation 6.29 can also be expressed in terms of the temporal frequency ν. Given that $\omega = 2\pi\nu$, we have $d\omega/d\nu = 2\pi$ so that $d\omega = 2\pi\, d\nu$. Substituting this into Equation 6.29,

$$f(t) = \int_{\nu=-\infty}^{\infty} g(\nu)\, e^{i2\pi\nu t}\, d\nu \tag{6.31}$$

where, with a slight abuse of notation, we have redefined $g(2\pi\nu)$ as $g(\nu)$. Similarly, upon substituting $\omega = 2\pi\nu$, Equation 6.30 becomes

$$g(\nu) = \int_{t=-\infty}^{\infty} f(t)\, e^{-i2\pi\nu t}\, dt. \tag{6.32}$$

Again, because the product of the temporal frequency ν and time t is dimensionless, these are conjugate variables, as depicted in Figure 6.1.

Fourier Pair Constants. Fourier pairs can be written in a more general form. For example, the Fourier pair defined in Equations 6.29 and 6.30 can be written as

$$f(t) = K_f \int_{\omega=-\infty}^{\infty} g(\omega)\, e^{i\omega t}\, d\omega, \tag{6.33}$$

$$g(\omega) = K_g \int_{t=-\infty}^{\infty} f(t)\, e^{-i\omega t}\, dt. \tag{6.34}$$

It is important to note that even though the value of each of the constants K_f and K_g is arbitrary, the value of their product $K_f K_g$ is not. For example, if we substitute $g(\omega)$ from Equation 6.34 into Equation 6.33, we get

$$f(t) \quad = \quad K_f K_g \int_{\omega=-\infty}^{\infty} \left[\int_{t=-\infty}^{\infty} f(t) e^{-i\omega t} dt \right] e^{i\omega t} d\omega. \qquad (6.35)$$

In order to maintain equality on both sides, if K_f is made smaller then K_g must be made larger, and vice versa. This has important consequences for quantum mechanics (see Section 8.4).

Symmetric Fourier Pairs. Given that the constraint on the pre-factors K_f and K_g is defined in terms their product $K_f K_g$, the Fourier pair in Equations 6.29 and 6.30 is often written in the symmetric form

$$f(t) \quad = \quad \frac{1}{\sqrt{2\pi}} \int_{\omega=-\infty}^{\infty} g(\omega) e^{i\omega t} d\omega, \qquad (6.36)$$

$$g(\omega) \quad = \quad \frac{1}{\sqrt{2\pi}} \int_{t=-\infty}^{\infty} f(t) e^{-i\omega t} dt. \qquad (6.37)$$

Temporal Frequency and Angular Frequency. It is not always obvious whether the sinusoids in a Fourier transform are measured in units of cycles/s or radians/s. For complex Fourier transforms, the way to spot the difference is to look at the exponent. If this has the form $i2\pi\nu t$ then ν is a temporal frequency, measured in cycles/s or Hz (e.g. Equation 6.31). If the exponent has the form $i\omega t$ then ω is an angular frequency, measured in units of radians/s (e.g. Equation 6.29).

6.4. Summary

To augment the summary in Chapter 2, the Fourier transform of $f(t)$ can be defined in terms of three different but equivalent sets of parameter pairs. Each pair of parameters determines the amplitude and phase of a sinusoid at a single frequency ω_n. These three pairs of parameters are:

1. the amplitude A_n and phase θ'_n of a sinusoid;
2. the amplitude C_n of a cosine and amplitude D_n of a sine, which together define a sinusoid with amplitude $A_n = \sqrt{C_n^2 + D_n^2}$ and phase $\theta'_n = \arctan(D_n/C_n)$;
3. the complex coefficients $\hat{A}_n = (C_n - iD_n)/2$ and $\hat{A}_{-n} = (C_n + iD_n)/2$, which together define a sinusoid with amplitude $A_n = |\hat{A}_n| + |\hat{A}_{-n}|$ and phase $\theta'_n = \arctan(D_n/C_n)$.

In principle, the function $f(t)$ can be real, imaginary or complex. However, because we assume that the signal $f(t)$ is a physical quantity, it is taken to be a real function in this book.

Chapter 7

Properties of Fourier Transforms

7.1. Introduction

Consider a *real function* $f(t)$ of time t and its Fourier transform, which is a complex function $g(\omega)$ of angular frequency ω. We represent a function and its Fourier transform with the standard notation

$$f(t) \quad \Leftrightarrow \quad g(\omega), \tag{7.1}$$

where $f(t)$ and $g(\omega)$ are defined in Equations 6.29 and 6.30. Expressed in terms of frequency,

$$f(t) \quad \Leftrightarrow \quad g(\nu), \tag{7.2}$$

where $f(t)$ and $g(\nu)$ are defined in Equations 6.31 and 6.32.

7.2. Dirichlet Conditions

Only functions that meet certain criteria, known as *Dirichlet conditions*, can be Fourier transformed. Fortunately, almost all natural phenomena give rise to functions that satisfy these criteria. For this reason, we do not list the Dirichlet conditions here; the interested reader can find them in more advanced texts.

7.3. Conjugate Variables

The key idea that underpins Fourier transforms is to express the value of a function $f(t)$ at any time t as the sum of the amplitudes of sinusoids at that time, where the amplitude of the sinusoid with frequency ν is $g(\nu)$. Crucially, the value of the function f is specified by a variable t with units of seconds s, whereas the value of the function g is specified by a variable ν with units of $1/s$. The product of t and ν, which constitute a Fourier pair (Section 6.3), is therefore dimensionless (because $t \times 1/t$

has no physical dimensions); thus t and ν are referred to as *conjugate variables*. The variables t and ω are also conjugate because the factor 2π in $\omega = 2\pi\nu$ means that the physical units of ω are also $1/s$, and therefore $t \times \omega$ has no physical dimensions.

7.4. The Sampling Theorem and Aliasing

Before we begin, it is worth noting that a *theorem* is simply a mathematical statement that has been proved to be true. When a signal is recorded for digital media, it is sampled at a particular rate. For example, music is typically sampled at the rate of $R = 40{,}000$ times per second ($40\,\text{kHz}$), which means it is sampled every $\Delta t = 1/R = 25$ microseconds. An obvious question is: what is the highest frequency that can be recovered from a signal if it is sampled at the rate of R samples/s?

The answer is $\nu = R/2$ Hz, which is known as the *Nyquist frequency*, as shown in Figure 7.1. Therefore, the highest frequency that can be detected in music which has been sampled at 40,000 samples/s is $20\,\text{kHz}$ (which is the highest frequency that can be heard by humans). This is called the *Nyquist–Shannon sampling theorem*.

Aliasing. Aliasing occurs when a high-frequency sinusoid masquerades as a low-frequency sinusoid, which is the result of *under-sampling* a signal. If a function is sampled at the rate of R samples/s (i.e. every $\Delta t = 1/R$ seconds) then the highest frequency that can be detected is the Nyquist frequency $\nu_{\text{Nyq}} = R/2$ Hz. However, if the sampled signal contains values from sinusoids with frequencies $\nu > \nu_{\text{Nyq}}$, these can fool a naive sampling strategy into believing that there are sinusoids present at frequencies below ν_{Nyq}, even if no such frequencies exist (see Figure 7.2). In practice, this can be avoided by filtering a signal before it is digitized to remove any frequencies above the Nyquist frequency.

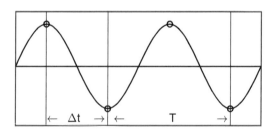

Figure 7.1: If a function is sampled every Δt seconds then the shortest period that can be represented is $T = 2\Delta t$, as shown here. Therefore, the Nyquist frequency (the highest frequency that can be represented) is $\nu_{\text{Nyq}} = 1/T = 1/(2\Delta t)$ Hz. Sampled values are marked with circles.

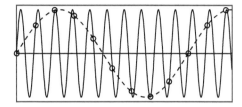

Figure 7.2: Aliasing. The interval between consecutive samples (circles) yields identical values for both the high- and the low-frequency sinusoidal functions. Consequently, the values sampled on the high-frequency sinusoid (solid curve) effectively impersonate a low-frequency sinusoid (dashed curve).

7.5. How Many Fourier Components?

Given an ordered set of N data points $x = \{x_1, \ldots, x_N\}$, how many components does the Fourier transform produce? The set x could be a temporal sequence of sound amplitudes or a sequence of pixel values from an image; here we will assume that x consists of sound values.

If the N sampled values span an interval of T seconds then the sampling rate is $R = N/T$ samples/s, so the Nyquist frequency is

$$\nu_{\text{Nyq}} = R/2 \tag{7.3}$$
$$= N/(2T) \text{ Hz.} \tag{7.4}$$

The longest sinusoid that will fit into an interval of T seconds has a period of T seconds, which means that the lowest nonzero (i.e. fundamental) frequency is

$$\nu_{\text{min}} = 1/T \text{ Hz.} \tag{7.5}$$

Because the Fourier transform returns frequencies that are integer multiples of ν_{min} up to ν_{Nyq}, the number of Fourier components is

$$\frac{\nu_{\text{Nyq}}}{\nu_{\text{min}}} = \frac{N/(2T)}{1/T} \tag{7.6}$$

$$= N/2, \tag{7.7}$$

so the number of Fourier components is half the number of sampled values. Of course, this is true whether the samples are from a sound, an image, or any physical quantity. For example, if x is a row of image pixel values then the sinusoids have frequencies defined in terms of cycles per metre, which is a measure of *spatial frequency* (see Sections 7.15 and 8.1).

7.6. The Addition Theorem

The sum of two functions has a Fourier transform that equals the sum of the Fourier transforms of the individual functions:

$$f_1(t) + f_2(t) \quad \Leftrightarrow \quad g_1(\omega) + g_2(\omega). \tag{7.8}$$

7.7. The Shift Theorem

If a function $f(t)$ is shifted by an amount Δt then its Fourier transform gets multiplied by $e^{i\omega\Delta t}$:

$$f(t + \Delta t) \quad \Leftrightarrow \quad g(\omega)\, e^{i\omega\Delta t}. \tag{7.9}$$

7.8. Parseval's Theorem

Parseval's theorem states that power is conserved when a signal $f(t)$ is transformed into Fourier coefficients $\hat{A}_\nu$. Specifically,

$$\int_{t=-\infty}^{\infty} |f(t)|^2\, dt \quad = \quad \int_{\nu=-\infty}^{\infty} |\hat{A}_\nu|^2\, d\nu. \tag{7.10}$$

In words, the power of a signal $f(t)$ is the sum of the powers $|\hat{A}_\nu|^2$ at different frequencies.

From the perspective of physics, it seems obvious that Parseval's theorem must be true. For example, when light passes through an aperture, the resultant diffraction pattern on a screen consists of the same light that passed through the aperture. As we shall see, when light passes through an aperture, the diffraction pattern on a screen results from a Fourier transform of the shape of the aperture. The conservation of energy ensures that the energy arriving at the aperture must be the same as the energy arriving at the screen, which is why Parseval's theorem is reasonably obvious.

7.9. The Convolution Theorem

When a physical quantity x is measured, the measuring device inevitably 'smears out' x over time and space. For example, the lens in a camera distorts the image measured at the light-sensitive image plane; in this case, the distortion is characterised by the *point spread function* (PSF) of the lens. In effect, it is as if a smearing function (e.g. a PSF) is swept over the scene in front of the camera (which can be interpreted as a platonic, or 'perfect version' of an, image), and the result is the recorded image.

For simplicity, we consider the intensity $f_1(t)$ along a single row of pixels in a platonic image, where t indicates position. For such one-dimensional data, the smearing function is typically called the *resolution function* or the *impulse response function*.

The convolution of $f_1(t)$ with a smearing function $f_2(t)$ produces the measured image

$$C(t) \quad = \quad f_1(t) * f_2(t) \tag{7.11}$$

$$= \quad \int_{t'=-\infty}^{\infty} f_1(t') f_2(t - t') \, dt', \tag{7.12}$$

where the symbol $*$ represents the *convolution operator*. The *convolution theorem* states that the convolution of $f_1(t)$ with $f_2(t)$ can be obtained from the inner product of their Fourier transforms,

$$f_1(t) * f_2(t) \quad \Leftrightarrow \quad g_1(\nu) \times g_2(\nu), \tag{7.13}$$

where the product $g_1(\nu) \times g_2(\nu)$ involves multiplying $g_1(\nu)$ at frequency ν by the corresponding value of $g_2(\nu)$ at frequency ν. Convolution is computationally fairly expensive, whereas the Fourier transform is efficient, so it often makes sense to take advantage of the convolution theorem. More importantly, the convolution theorem provides a powerful tool, called *deconvolution*, for recovering the underlying signal from a 'smeared out' temporal sequence of measurements (see Section 8.2).

7.10. Counting Fourier Parameters

Here we compare the number of values required to sample a signal $f(t)$ every Δt seconds with the number of parameters required to represent the Fourier transform of that signal.

First, we find the number of parameters required to represent the Fourier transform of $f(t)$. If $f(t)$ lasts for T seconds then the lowest nonzero frequency in $f(t)$ is

$$\nu_{\min} \quad = \quad 1/T. \tag{7.14}$$

If $f(t)$ is sampled every Δt seconds then the Nyquist frequency is

$$\nu_{\max} \quad = \quad \frac{1}{2\Delta t}. \tag{7.15}$$

The frequencies of Fourier components are integer multiples of $\nu_{\min}$, so the number of frequencies between $\nu_{\min}$ and $\nu_{\max}$ is

$$N_{\text{sinusoids}} = \frac{\nu_{\max}}{\nu_{\min}} \tag{7.16}$$

$$= \frac{T}{2\Delta t}. \tag{7.17}$$

But each component is specified by two numbers, the amplitude A and the phase θ' (or the coefficients C and D), so the total number of parameters is

$$N_{\text{Fourier}} = 2 \times N_{\text{sinusoids}} \tag{7.18}$$

$$= T/\Delta t. \tag{7.19}$$

Next, we find the number of values required to represent $f(t)$ at a temporal resolution of Δt seconds. In order to record sample values of $f(t)$ every Δt seconds over a period of T seconds, the total number of values we need to store is

$$N_{\text{samples}} = T/\Delta t. \tag{7.20}$$

We can therefore conclude that there is no net saving, in terms of storage space, whether we choose to represent $f(t)$ as N_{samples} values or as N_{Fourier} Fourier parameters. This, in turn, suggests that the same amount of information is contained in both representations.

7.11. Fourier Transform of a Gaussian

The *Gaussian distribution* is a bell-shaped function with two parameters, the mean μ_x and the *standard deviation* σ_x, and takes the form

$$f(x) = \frac{1}{\sigma_x} \frac{1}{\sqrt{2\pi}} \times e^{-(x-\mu_x)^2/(2\sigma_x^2)}. \tag{7.21}$$

The standard deviation is a measure of the variability in x, which sets the width of the Gaussian distribution, as shown in Figure 7.3. Given M samples $x_1, \ldots, x_M$ of x, the standard deviation is

$$\sigma_x = \left(\frac{1}{M} \sum_{i=1}^{M} (x_i - \mu_x)^2 \right)^{1/2}. \tag{7.22}$$

To simplify notation, we assume $\mu_x = 0$, so that

$$f(x) \quad = \quad \frac{1}{\sigma_x} \frac{1}{\sqrt{2\pi}} \times e^{-x^2/(2\sigma_x^2)}. \tag{7.23}$$

The factor $1/(\sigma_x\sqrt{2\pi})$ ensures that $f(x)$ is a *normalised Gaussian*, which means that the area under the curve is 1, i.e.

$$\int_{x=-\infty}^{\infty} f(x)\,dx \quad = \quad 1. \tag{7.24}$$

This matters if we choose to treat $f(x)$ as a *probability distribution*.

The Fourier transform of $f(x)$ in terms of angular frequency ω is

$$g(\omega) \quad = \quad \frac{1}{\sqrt{2\pi}} \int_x f(x)\,e^{i\omega x}\,dx. \tag{7.25}$$

It turns out that the Fourier transform of a Gaussian is another Gaussian,

$$g(\omega) \quad = \quad \frac{1}{\sqrt{2\pi}} \times e^{-\omega^2\sigma_x^2/2}. \tag{7.26}$$

From the general form of a Gaussian in Equation 7.23, we see that the exponential part of the Gaussian in Equation 7.26 must satisfy

$$g(\omega) \quad \propto \quad e^{-\omega^2/(2\sigma_\omega^2)}. \tag{7.27}$$

Comparing this with Equation 7.26 implies

$$\frac{\omega^2\sigma_x^2}{2} \quad = \quad \frac{\omega^2}{2\sigma_\omega^2}, \tag{7.28}$$

and therefore $\sigma_x\,\sigma_\omega = 1$. In words, the Fourier transform of a normalised Gaussian function $f(x)$ with standard deviation σ_x is another Gaussian

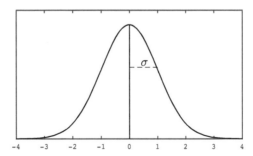

Figure 7.3: A Gaussian function with a mean of zero and standard deviation $\sigma = 1$, as indicated by the horizontal dashed line.

function $g(\omega)$ with standard deviation $\sigma_\omega = 1/\sigma_x$. Thus, as σ_x increases, σ_ω shrinks, and vice versa. More importantly, it can be shown that if $f(x)$ is not Gaussian then $\sigma_x \sigma_\omega > 1$. These results are enshrined as *Heisenberg's inequality*, which states that

$$\sigma_x \, \sigma_\omega \;\; \geq \;\; 1, \tag{7.29}$$

with equality only if $f(x)$ is Gaussian. This is the basis of *Heisenberg's uncertainty principle* (see Section 8.4).

7.12. Trading Frequency for Time

The result obtained with a Gaussian distribution demonstrates a general feature of Fourier transforms: if a signal is localised in time then it is spread out in frequency, and vice versa. This means that a signal that is concentrated within a short span of time translates to a set of Fourier components that are widely distributed in frequency. For example, the short signal in Figure 7.4a has a broad amplitude spectrum as shown in Figure 7.4b. Similarly, the signal on the right end panel of Figure 3.1 is like a pulse, but we can see that the amplitude spectrum of its Fourier transform (displayed along the frequency axis) consists of a broad range of frequencies.

Conversely, a compact amplitude spectrum translates to a signal that is evenly distributed over time, such as a single, constant sinusoidal tone. In the case of a pure sinusoidal tone with frequency 2 Hz, as in Figure 7.5a, the amplitude spectrum of its Fourier transform is a single spike at 2 Hz, as shown in Figure 7.5b.

More generally, when expressed in terms of Fourier pairs (Section 6.3), time and temporal frequency can be replaced with space and spatial frequency or any other pair of conjugate physical variables. In all of these cases, if one member of a pair of conjugate variables is localised then the other is spread out.

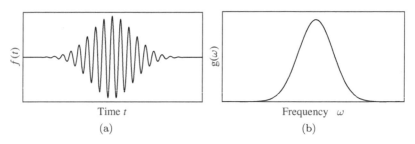

Figure 7.4: (a) A brief signal localised in time. (b) The amplitude spectrum of the Fourier transform of (a) is a Gaussian function.

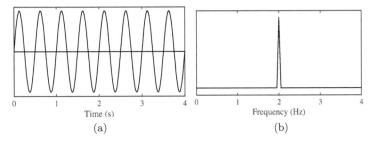

Figure 7.5: (a) A pure sinusoid with frequency 2 Hz. (b) The amplitude spectrum of the Fourier transform of (a) consists of a single spike at 2 Hz.

7.13. Information Theory and Fourier Transforms

Information theory was invented by Claude Shannon and first published in 1948. The theorems of information theory are so important that they deserve to be regarded as the *laws of information*.

Information is measured in *bits*, where one bit is the amount of information required to choose between two equally probable alternatives. More generally, if a variable x adopts the value x_i with probability $p(x_i)$ then the amount of information conveyed by the value x_i is

$$h(x_i) \quad = \quad \log_2 \frac{1}{p(x_i)} \text{ bits,} \tag{7.30}$$

where use of logarithms to base 2 ensures that information is measured in bits. As we should expect, values of x that are less probable (i.e. more surprising) convey more information than more probable values.

The *Shannon information* or *Shannon entropy* of the distribution $p(x)$ is the average amount of information conveyed by a value x. Specifically, a variable x that adopts each of n possible values $x_1, \ldots, x_n$ with corresponding probabilities $p(x_1), \ldots, p(x_n)$ has a *Shannon entropy* of

$$H(x) \quad = \quad \sum_{i=1}^{n} p(x_i) h(x_i) \tag{7.31}$$

$$= \quad \sum_{i=1}^{n} p(x_i) \log_2 \frac{1}{p(x_i)} \text{ bits.} \tag{7.32}$$

An extremely useful property of Fourier analysis is that when applied to *any* variable, the resultant Fourier components are *uncorrelated*. More importantly, if Gaussian variables are uncorrelated then they are also *mutually independent*.

The distinction between uncorrelatedness and independence is subtle. For our purposes, we only need to know that if two variables are uncorrelated then the value of one variable can still provide information about the value of the other variable, whereas independence means that the value of one variable cannot possibly provide any information on the value of the other variable.

In practice, it is not easy to estimate the entropy of a given physical quantity. However, because the Fourier components of a Gaussian variable are independent, the amount of information implicit in that variable can be estimated by adding up the information contained in its Fourier components. This, in turn, can be used to estimate the rate at which information is transmitted through a communication channel.

A communication channel can be a telephone line, a satellite link, or a neuron connecting the eye to the brain. The rate at which information about the channel input arrives at the channel output is the *mutual information* between input and output, and the maximum rate at which information can be transmitted through a channel is the *channel capacity*. In practice, all channels add *noise* to the input signal, which reduces the channel capacity.

Consider a variable $y = x + \eta$, which is the sum of a Gaussian signal x with *variance* S and Gaussian noise η with variance N, where variance is a measure of variability (S and N here are standard notation in this context). For example, given a variable x with mean μ_x, its variance is

$$S \;=\; \frac{1}{M} \sum_{i=1}^{M} (x_i - \mu_x)^2, \tag{7.33}$$

which is the square of the standard deviation (Equation 7.22). If the highest frequency in y is W Hz, and if values of x are transmitted at the Nyquist rate of $2W$ Hz, then the channel capacity is

$$C \;=\; W \log\left(1 + \frac{S}{N}\right) \text{ bits/s.} \tag{7.34}$$

After Fourier analysis, if the signal power of the Fourier component $x(\nu)$ is $S(\nu)$ and the noise power of the component $\eta(\nu)$ is $N(\nu)$, the *signal-to-noise ratio* is $S(\nu)/N(\nu)$. The mutual information between x and y at frequency ν is therefore

$$I\big(x(\nu), y(\nu)\big) \;=\; \log\left(1 + \frac{S(\nu)}{N(\nu)}\right) \text{ bits/s.} \tag{7.35}$$

As mentioned earlier, the Fourier components of any Gaussian variable are mutually independent. Therefore, the mutual information between the Gaussian variables x and y can be obtained by integrating

$I(x(\nu), y(\nu))$ over frequency:

$$I(x, y) \quad = \quad \int_{\nu} I\big(x(\nu), y(\nu)\big) \, d\nu \quad \text{bits/s.} \tag{7.36}$$

To summarise, provided that the input, noise and outputs have Gaussian distributions, the rate at which information is transmitted through a channel is obtained by adding up the transmission rates at each frequency. Using these Gaussian assumptions, the total transmission rate equals the channel capacity (i.e. $I(x, y) = C$). If non-Gaussian distributions are used then the transmission rate will be less than the channel capacity.

7.14. The Fast Fourier Transform

A critical factor in the practical utility of any algorithm is its speed of execution. This is typically described in terms of *time complexity*, which describes how quickly the time required increases with the size of the problem, measured as the number of data samples N used with the Fourier transform. Given N data points sampled from a signal, we can work out the number of numerical operations as follows.

The number of numerical operations required by the discrete Fourier transform (DFT) to obtain the coefficient C_n at one frequency ω_n is N, and these operations are repeated for every frequency up to the Nyquist frequency (see Equations 4.29 and 4.30). Given that the Nyquist frequency is $f_{\text{Nyq}} = N/2$, the number of frequencies is $N/2$, so the total number of numerical operations required to obtain all $N/2$ Fourier coefficients C_n is proportional to N^2. Of course, this also applies to D_n, but we are not really interested in the absolute number of operations. Instead, we want to know how quickly the number of operations increases with N. In the case of the discrete Fourier transform, this is written as

$$M_{\text{DFT}} \quad = \quad O(N^2), \tag{7.37}$$

where the capital letter O represents the *order* of the time complexity function. A brief examination of the main loop in the discrete Fourier transform in Appendix A or B will confirm this.

In contrast, the *fast Fourier transform* (FFT) uses a recursive method to reduce the number of numerical operations, so that

$$M_{\text{FFT}} \quad = \quad O(N \log N). \tag{7.38}$$

For example, if $N = 4096$ then $M_{\text{DFT}} = 30$ million, whereas $M_{\text{FFT}} = 30,000$, so the FFT represents a 1000-fold saving in this case. A graph comparing M_{DFT} with M_{FFT} for $N = 1, \ldots, 50$ is shown in Figure 7.6.

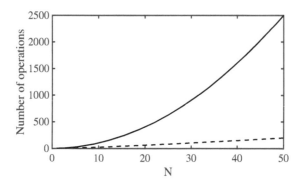

Figure 7.6: Comparison of the time complexity $M_{\text{DFT}} = O(N^2)$ (solid curve) of the Fourier transform with the time complexity $M_{\text{FFT}} = O(N \log N)$ (dashed curve) of the fast Fourier transform, for $N = 1, \ldots, 50$. For simplicity, constants have been set to 1 here, so that the plotted curves are $M_{\text{DFT}} = N^2$ and $M_{\text{FFT}} = N \log N$.

The FFT was published in 1965 by Cooley and Tukey, which is why it is also called the *Cooley–Tukey algorithm*. It had a substantial impact on crystallography, where the computation of Fourier transforms represented a major bottleneck (see Section 8.5). More recently, audio and video streaming applications depend critically on the ability to perform Fourier transforms quickly (see Section 8.1).

7.15. Two-Dimensional Fourier Transforms

Just as a temporal signal consists of an ordered sequence of values, so an image consists of a two-dimensional array of intensity values. Thus, in translating from temporal signals to images, we can replace the time index t with two spatial indices x and y.

By analogy with the one-dimensional inverse Fourier transform of a temporal signal in Equation 6.31 (repeated here),

$$f(t) \quad = \quad \int_{\nu=-\infty}^{\infty} g(\nu)\, e^{i2\pi\nu t}\, d\nu, \tag{7.39}$$

the two-dimensional inverse Fourier transform is

$$f(x,y) \quad = \quad \int_{v=-\infty}^{\infty} \int_{u=-\infty}^{\infty} g(u,v)\, e^{i2\pi(ux+vy)}\, du\, dv. \tag{7.40}$$

Just as the number of temporal periods (cycles) executed by the sinusoid with frequency ν at time t is νt in Equation 7.39, so the number of spatial periods (wavelengths) at location (x,y) in the array is $ux + vy$. And just as $2\pi\nu t$ is the phase of a sinusoidal frequency at time t, so

$2\pi(ux + vy)$ represents the phase of a spatial sinusoidal frequency at location (x, y).

Consider the picture in Figure 7.7a. What would it mean to have a Fourier transform of this picture? Well, we would like to know the amplitude of each spatial frequency along each orientation, as if a single one-dimensional Fourier transform had been obtained along every orientation in the image. Of course, it makes sense to store the results of these multiple one-dimensional Fourier transforms in a two-dimensional array, and this is the two-dimensional Fourier transform. In fact, the Fourier transform of an image requires two two-dimensional arrays, one for the amplitudes and one for the phases, as shown in Figure 7.7b and c.

Polar Coordinates. A two-dimensional Fourier transform is most conveniently described using *polar coordinates*. Instead of specifying the location in an image as (x, y), we can use the distance r from the origin and the direction θ, where

$$
\begin{aligned}
r &= \sqrt{x^2 + y^2}, \\
\theta &= \arctan(x/y).
\end{aligned}
\tag{7.41}
$$

Conversely, we can express x and y in terms of r and θ:

$$
\begin{aligned}
x &= r\cos\theta, \\
y &= r\sin\theta.
\end{aligned}
\tag{7.42}
$$

Similarly, we can express the location (u, v) in a two-dimensional Fourier transform in terms of the polar coordinates (ρ, ϕ):

$$
\begin{aligned}
u &= \rho\cos\phi, \\
v &= \rho\sin\phi.
\end{aligned}
\tag{7.43}
$$

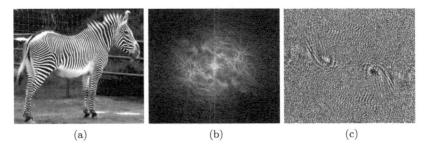

(a)   (b)   (c)

Figure 7.7: (a) Image. (b) Amplitude spectrum, where the spatial frequency increases with distance from the centre. (c) Phase spectrum, where swirls reflect the change in orientation and spatial frequency across the zebra's body.

Here, the distance ρ from the origin increases in proportion to spatial frequency, and the angle ϕ represents orientation. Substituting Equations 7.42 and 7.43 into Equation 7.40 gives

$$f(r,\theta) = \int_{\phi=0}^{2\pi} \int_{\rho=0}^{\infty} g(\rho,\phi)\, e^{i2\pi((\rho\cos\phi)(r\cos\theta)+(\rho\sin\phi)(r\sin\theta))}\, \rho\, d\rho\, d\phi,$$

(7.44)

which simplifies to

$$f(r,\theta) = \int_{\phi=0}^{2\pi} \int_{\rho=0}^{\infty} g(\rho,\phi)\, e^{i2\pi\rho r\cos(\theta-\phi)}\, \rho\, d\rho\, d\phi. \quad (7.45)$$

The two-dimensional Fourier transform of the image is

$$g(u,v) = \int_{y=-\infty}^{\infty} \int_{x=-\infty}^{\infty} f(x,y)\, e^{-i2\pi(ux+vy)}\, dx\, dy, \quad (7.46)$$

which can also be expressed in polar-coordinate form

$$g(\rho,\phi) = \int_{\theta=0}^{2\pi} \int_{r=0}^{\infty} f(r,\theta)\, e^{-i2\pi\rho r\cos(\theta-\phi)}\, r\, dr\, d\theta. \quad (7.47)$$

As demonstrated in Figure 7.7, the value of the amplitude spectrum at (ρ, ϕ) represents the amplitude of the spatial frequency ρ along the orientation $\phi + \pi/2$ in the image. Similarly, the value of the phase spectrum at (ρ, ϕ) represents the phase of the spatial frequency ρ along the orientation $\phi + \pi/2$ in the image.

Finally, notice that the two-dimensional Fourier transform is plotted with its origin at the centre of each array in Figure 7.7b and c, and that each array is symmetric with respect to the origin. This symmetry reflects the fact that the two-dimensional Fourier transform can be regarded as a series of one-dimensional Fourier transforms, where each one-dimensional Fourier transform is associated with a single line that passes through the centre of the image.

Chapter 8

Applications

8.1. Satellite TVs, MP3s and All That

Almost any form of digital communication depends on Fourier analysis. This is because Fourier analysis enables most visual and auditory information that humans cannot perceive to be discarded before data are transmitted. For example, the human ear cannot detect sound at frequencies above 20 kHz, so there is no point in wasting valuable bandwidth on transmitting data above 20 kHz. Similarly, the human visual system has limited ability to see beyond a certain level of detail, or to perceive motion above certain speeds. Also, humans are particularly bad at discriminating colours.

All of these human limitations mean that a large proportion of data recorded by cameras and microphones can be discarded. And of course, the Fourier transform makes it easy to select unwanted spatial frequencies in images, or unwanted temporal signals in sound, and throw them away. In practice, the precise form of Fourier transform used is usually a cut-down version (e.g. the *discrete cosine transform*), but the principle remains the same.

Compression that throws away information is called *lossy*, whereas compression that preserves all information is called *lossless*. Most practical systems are lossy, but they do not appear to lose information because the discarded information is undetectable by humans, or at least unimportant for human perception of images and sounds.

A television camera records data at a rate of about 1.5 gigabits per second (Gb/s), where a gigabit is one billion (10^9 or 1000 million) binary digits. This figure of 1.5 Gb/s results from the fact that TVs typically display 1920 elements horizontally and 1080 lines vertically, at a rate of 30 images per second. Each colour image displayed consists of 1920 × 1080 sets of three pixels (red, green and blue), so the total number of pixels is about 6 million ($3 \times 1920 \times 1080 = 6{,}220{,}800$).

Let's assume that the intensity of each pixel has a range of 0 to 255, which is represented by $\log_2(256) = 8$ binary digits, making a total of

49,766,400, or about 50 million, binary digits per image. Because a TV displays 30 images per second, this amounts to about 1500 million binary digits per second. In the world of computing this is confusingly quoted as 1500 megabits/s (i.e. 1500 million bits/s) because both 'bit' and 'binary digit' are used to refer to binary digits.

However, these 1500 million binary digits per second are transmitted through a channel (e.g. a satellite) that can carry only 19.2 million binary digits per second. In essence, this is done by discarding certain spatial and temporal frequencies and recoding the intensity/colour data to achieve an effective compression factor of about 78 ($\approx 1500/19.2$), so that it looks as if we can communicate 78 times more data than the channel capacity would suggest. In fact, the compression has to be a little better than this, because the stereo sound is also squeezed into the same channel. This is achieved by applying the cosine transform mentioned above to sound (where it is called MP3).

The standard methods used to remove spatial and temporal redundancy are collectively called MPEG (Moving Picture Expert Group). Although these methods are quite complex, they all rely heavily on a close relative of the Fourier transform, the *cosine transform*.

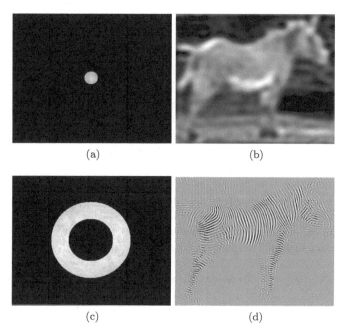

(a) (b)

(c) (d)

Figure 8.1: (a) The central region of the Fourier transform in Figure 7.7a encompasses low spatial frequencies. (b) The inverse transform of (a) yields a fuzzy zebra. (c) The annulus encompasses a band of spatial frequencies in Figure 7.7b. (d) The inverse transform of (c) corresponds to narrow stripes.

Compressing a Zebra. As an example of data compression, consider the image of a zebra and its Fourier transform in Figure 7.7. Each of them has $N_x = 308$ rows and $N_y = 362$ columns, making a total of $N = N_x N_y = 111{,}496$ elements. If we restrict the range of frequencies used to invert the Fourier transform to low spatial frequencies, as in Figure 8.1a, then the result is a fuzzy zebra, as shown in Figure 8.1b.

This is a rather extreme example of data compression, but the zebra is still recognisably a zebra. In this case, the data used to represent the fuzzy zebra are restricted to the first 15 frequencies (i.e. the disc in Figure 8.1a has a radius of $r = 15$). Each pixel within this disc corresponds to a Fourier component, each of which is specified by two numbers, the amplitude and phase (not shown). Therefore, the total amount of data represented within the disc is equal to twice its area, i.e. $2\pi r^2 = 2\pi \times 15^2$. But the symmetry of the Fourier transform means that each Fourier component is represented twice, so the total amount of non-redundant data is the area of the disc, $a = \pi \times 15^2 \approx 707$.

In contrast, the total amount of data storage required for the full Fourier transform of Figure 7.7a is $A = 2N_x N_y/2 = 111{,}496$. Therefore, the ratio of the amount of data required for the reduced Fourier transform in Figure 8.1b to the amount of data required for the full Fourier transform of Figure 7.7a is $a/A = \pi r^2/(N_x N_y) = 707/111{,}496 = 0.0063$.

Of course, we can use the Fourier transform to remove any range of spatial frequencies. For example, in Figure 8.1c all frequencies except those corresponding to radii between $r = 50$ and $r = 90$ have been removed from the amplitude spectrum in Figure 7.7b. Consequently, the inverse Fourier transform of Figure 8.1c displays Fourier components with spatial frequencies that correspond to stripes on the zebra. Similarly, any unwanted spatial or temporal frequencies (such as mains hum) can be selectively removed from data.

8.2. Deconvolution

Consider a measurement device with an impulse response function (IRF) $f_2(t)$. This means that if the system is exposed to an input signal $f_1(t)$, the response (i.e. the measured signal) results from the convolution of $f_1(t)$ with $f_2(t)$,

$$f(t) = f_1(t) * f_2(t), \tag{8.1}$$

as shown in Figure 8.2. The Fourier pairs are

$$f(t) \iff g(\nu) \quad \text{(measured signal)}, \tag{8.2}$$
$$f_1(t) \iff g_1(\nu) \quad \text{(signal)}, \tag{8.3}$$
$$f_2(t) \iff g_2(\nu) \quad \text{(IRF)}. \tag{8.4}$$

From Section 7.9, the Fourier transform of the measured signal $f(t)$ is

$$g(\nu) \;=\; g_1(\nu) \times g_2(\nu). \qquad (8.5)$$

Therefore, the Fourier transform of the signal $f_1(t)$ is

$$g_1(\nu) \;=\; \frac{g(\nu)}{g_2(\nu)}. \qquad (8.6)$$

Finally, the inverse Fourier transform of $g_1(\nu)$ provides an estimate $f_1'(t)$ of the signal $f_1(t)$, as in Figure 8.2.

In words, divide the Fourier transform $g(\nu)$ of the measured signal $f(t)$ by the Fourier transform $g_2(\nu)$ of the IRF $f_2(t)$ (for each frequency ν) to obtain the Fourier transform $g_1(\nu)$ of $f_1(t)$. The deconvolved signal $f_1'(t) \approx f_1(t)$ is obtained as the inverse Fourier transform of $g_1(\nu)$.

In practice, this procedure can be compromised by the presence of high-frequency measurement noise $\eta(t)$, which gets added to $f(t)$. The IRF $f_2(t)$ typically has low power at high frequencies (as in Figure 8.2), so the frequency-by-frequency ratio in Equation 8.6 has small values

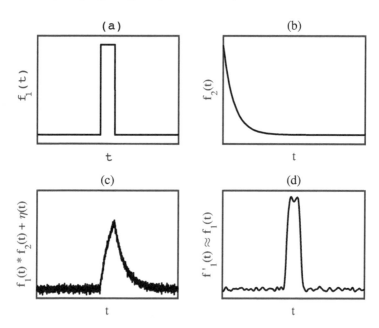

Figure 8.2: Schematic example of deconvolution.
(a) Original signal $f_1(t)$.
(b) Impulse response function $f_2(t)$ of measuring device.
(c) Measured signal $f(t) = f_1(t) * f_2(t) + \eta(t)$, where $\eta =$ high-frequency noise.
(d) Result of applying deconvolution to (de-noised version of) the measured signal in (c) to obtain the estimate $f_1'(t) \approx f_1(t)$.

in the denominator $g_2(\nu)$ at high frequencies, which artificially boosts high frequencies in the Fourier transform $g_1(\nu)$ and therefore in $f_1'(t)$. Various techniques exist to reduce the effects of noise, and the *maximum entropy (maxent)* method seems to be particularly effective.

8.3. Fraunhofer Diffraction

Fraunhofer diffraction is a beautiful example of how light passing through a small aperture provides a Fourier transform of the shape of that aperture. In effect, the aperture acts as an analogue computer that implements Fourier analysis. This is most easily understood when the aperture is a slit, as shown in Figure 8.3.

To set the scene, when light passes through a slit, it forms a characteristic diffraction pattern on a screen placed beyond the slit. The light source lies along the horizontal z-axis, at a distance sufficiently large that all points within the slit receive light waves with the same phase. Consequently, in accordance with *Huygens' wave theory of light*, we assume that the slit acts as a source of *coherent light*, such that every point within the slit generates a continuous stream of *wavelets*

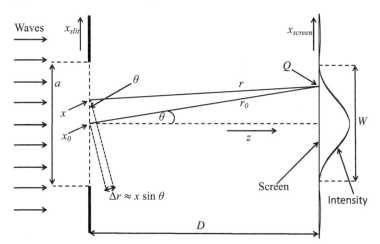

Figure 8.3: Fraunhofer diffraction. The phase of a wavelet w_0 from x_0 when it reaches Q is $\alpha_0 = 2\pi r_0/\lambda$, where r_0 is the path length from x_0 to Q. Similarly, the phase of a wavelet w from x when it reaches Q is $\alpha = 2\pi r/\lambda$. The combined contribution of w_0 and w to the intensity at Q depends on the (constant) phase difference $\Delta\alpha = \alpha - \alpha_0$ at Q, which depends on $\Delta r = r - r_0$, which in turn depends on $\Delta x = |x - x_0|$. For example, if $\Delta r = \lambda/2$ then w_0 and w cancel at Q, so the contribution of w_0 and w to the intensity at Q is zero; and this is true for every pair of points separated by Δx. A key assumption is that the slit-to-screen distance D is large, so the lines labelled r and r_0 are almost parallel and therefore $\Delta r \approx x \sin\theta$.

that have the same phase and frequency. The intensity at a point x within the slit is the square of the *field strength*, which is

$$E(x) \quad = \quad E_0 e^{2\pi \nu t}, \tag{8.7}$$

where ν is the frequency of light and E_0 is a constant.

The intensity at a point Q on the screen depends on whether the waves arriving at Q from different parts of the slit interfere with each other *constructively* or *destructively*. This, in turn, depends on the phases of waves arriving at Q, which depend on the distances or *path lengths* from different points within the slit to Q.

In essence, the contribution to the intensity at Q from any two points x and x_0 within the slit depends on the difference between the phases of waves from x and x_0 when those waves reach Q. Because this phase difference is constant over time for x and x_0, the contribution from x and x_0 to Q is also constant over time. And, because the slit-to-screen distance is large in relation to the slit width, the phase difference at Q is the same for every pair of points that are separated by the same distance $\Delta x = |x - x_0|$. Crucially, this contribution is largest if there is no phase difference, and decreases as the phase difference increases, as shown in Figure 8.4. Consequently, the intensity at Q, which lies in

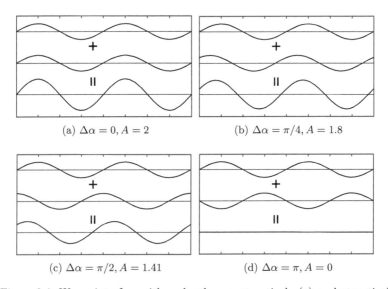

(a) $\Delta \alpha = 0, A = 2$ (b) $\Delta \alpha = \pi/4, A = 1.8$

(c) $\Delta \alpha = \pi/2, A = 1.41$ (d) $\Delta \alpha = \pi, A = 0$

Figure 8.4: Waves interfere with each other constructively (a) or destructively (b, c, d). In each panel, the top two sinusoids, each with unit amplitude, are added to yield the sinusoid at the bottom. The phase difference $\Delta \alpha$ between the top two sinusoids and the amplitude A of the summed sinusoid is shown below each panel.

direction θ, is the sum of contributions from all pairs of points separated by the same distance Δx within the slit.

More formally, consider a slit with its mid-point at x_0. A wavelet w_0 that originates at x_0 has a phase of α_0 at Q, which depends on the path length from x_0 to Q,

$$r_0 \quad = \quad \text{distance from } x_0 \text{ to } Q. \tag{8.8}$$

Similarly, a wavelet w that originates at point x within the slit has a phase α at Q that depends on the path length

$$r \quad = \quad \text{distance from } x \text{ to } Q. \tag{8.9}$$

Notice that, because wavelets from all points in the slit have the same frequency and phase, the phase difference

$$\Delta \alpha \quad = \quad \alpha - \alpha_0 \tag{8.10}$$

at a single point Q on the screen is the same at all times. This matters because the phase difference at Q determines the intensity at Q.

For example, if $\Delta \alpha = \pi$ then the wavelets from x_0 and x have opposite phases at Q, so they cancel each other. Consequently, the contribution of wavelets from x_0 and x to the field strength at Q is zero at all times. As we shall see, it is the value of $\Delta \alpha$ that determines the contribution of wavelets from x_0 and x to the field strength at Q. And, because $\Delta \alpha$ varies systematically with direction θ, it is ultimately θ that determines the contribution of wavelets from x_0 and x to the field strength at Q.

Interlude: Why Only the Phase Difference Matters. At this juncture, we take a small diversion to show why it is the phase difference $\Delta \alpha$ at Q that matters, and why we can ignore the variation of phase over time.

Consider a wavelet w that originates at x. The phase α of this wavelet at the time it reaches Q depends on the phase at x plus a constant phase increment due to the path length r from x to Q. For light with frequency ν Hz, the time-varying phase at x is

$$\alpha(x) \quad = \quad 2\pi\nu t. \tag{8.11}$$

When the phase of wavelet w at x is zero, the phase at Q is

$$\alpha \quad = \quad 2\pi r/\lambda. \tag{8.12}$$

Therefore, the time-varying phase at Q is

$$\alpha \quad = \quad 2\pi r/\lambda + 2\pi\nu t. \tag{8.13}$$

Now consider a wavelet w_0 that originates at x_0. If the path length from x_0 to Q is r_0 then the time-varying phase of w_0 at Q is

$$\alpha_0 = 2\pi r_0/\lambda + 2\pi\nu t. \tag{8.14}$$

Already, we can see that the variation of phase over time has no impact on the phase difference at Q because

$$\begin{aligned} \Delta\alpha &= \alpha - \alpha_0 \\ &= (2\pi r/\lambda + 2\pi\nu t) - (2\pi r_0/\lambda + 2\pi\nu t), \end{aligned} \tag{8.15}$$

with the time-varying phases cancelling, so that

$$\begin{aligned} \Delta\alpha &= 2\pi r/\lambda - 2\pi r_0/\lambda \\ &= 2\pi(r - r_0)/\lambda \\ &= 2\pi\Delta r/\lambda, \end{aligned} \tag{8.16}$$

where $\Delta r = r - r_0$ is the difference in path lengths from x and x_0 to Q.

Because we now know that the time-varying components of the contributions due to the wavelets from x and x_0 will cancel, we can safely drop the term $2\pi\nu t$ from now on. Accordingly, the phase of w_0 at Q is

$$\alpha_0 = 2\pi r_0/\lambda. \tag{8.17}$$

The contribution of w_0 to the field strength at Q is

$$g(\alpha_0, x) = E_0\, f(x)\, e^{i\alpha_0}, \tag{8.18}$$

where $f(x)$ is the *aperture function*, which is the wavelet amplitude at x. The term E_0 is a constant, which we set to $E_0 = 1$, so that

$$g(\alpha_0, x) = f(x)\, e^{i\alpha_0}. \tag{8.19}$$

If the aperture is a slit of width a then $f(x)$ is the *rectangular function*

$$f(x) = \begin{cases} 1 & \text{if } -a/2 < x < a/2, \\ 0 & \text{otherwise,} \end{cases} \tag{8.20}$$

as shown in Figures 8.3 and 8.5a. Similarly, assuming the phase of wavelet w at x to be zero, its phase at Q is

$$\alpha = 2\pi r/\lambda, \tag{8.21}$$

so the contribution of w to the field strength at Q is

$$g(\alpha, x) \;=\; f(x)\, e^{i\alpha}. \tag{8.22}$$

Using Equation 8.21, this can be expressed as a function of r and x,

$$g(r, x) \;=\; f(x)\, e^{i2\pi r/\lambda}. \tag{8.23}$$

However, we can express the path length r as

$$r \;=\; r_0 - \Delta r, \tag{8.24}$$

where Δr is the difference between the path lengths from x_0 and x to Q. Provided the slit-to-screen distance is large in relation to the slit width, the lines labelled r and r_0 in Figure 8.3 are almost parallel, so that

$$\Delta r \;\approx\; x \sin\theta \tag{8.25}$$

and therefore

$$r \;\approx\; r_0 - x \sin\theta. \tag{8.26}$$

By substituting this into Equation 8.23, the contribution to the field strength from x in the direction θ can be expressed as a function of x and θ,

$$\begin{aligned} g(\theta, x) \;&=\; f(x)\, e^{i2\pi(r_0 - x\sin\theta)/\lambda} \\ &=\; f(x)\, e^{i2\pi r_0/\lambda}\, e^{-i2\pi x(\sin\theta)/\lambda}. \end{aligned} \tag{8.27}$$

It is worth noting that in the exponent we have

$$2\pi(r_0 - x\sin\theta)/\lambda = 2\pi r_0/\lambda - 2\pi x(\sin\theta)/\lambda = \alpha_0 - \alpha, \tag{8.28}$$

which is the phase difference $\Delta\alpha$ at Q of the wavelets w_0 and w. This matters because Equation 8.28 states that this difference varies systematically with the direction θ and hence with screen position.

We can obtain the total field strength $g(\theta)$ in the direction θ by integrating the bivariate function $g(\theta, x)$ over position x within the slit,

$$g(\theta) \;=\; \int_x g(\theta, x)\, dx. \tag{8.29}$$

Substituting Equation 8.27, we obtain

$$g(\theta) \;=\; e^{i2\pi r_0/\lambda} \int_x f(x)\, e^{-i2\pi x(\sin\theta)/\lambda}\, dx, \tag{8.30}$$

which can be recognised as the Fourier transform of something — but of what?

Comparison with Equation 6.32 implies that the conjugate variables are x and

$$u = (\sin\theta)/\lambda, \tag{8.31}$$

so Equation 8.30 becomes

$$g(u) = e^{i2\pi r_0/\lambda} \int_x f(x)\, e^{-i2\pi xu}\, dx. \tag{8.32}$$

Therefore, $g(u)$ is the Fourier transform of $f(x)$, where $f(x)$ is the aperture function. In other words, the field strength at the screen is the Fourier transform of the aperture function. Using the rectangular function defined in Equation 8.20, Equation 8.32 evaluates to

$$g(u) = \kappa\, \frac{\sin(\pi a(\sin\theta)/\lambda)}{\pi a(\sin\theta)/\lambda}, \tag{8.33}$$

$$= \kappa \sin(\pi au)/(\pi au) \tag{8.34}$$

$$= \kappa \operatorname{sinc}(\pi au), \tag{8.35}$$

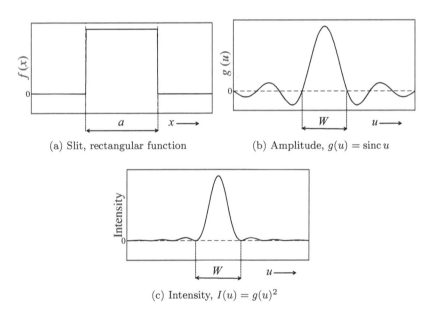

(a) Slit, rectangular function

(b) Amplitude, $g(u) = \operatorname{sinc} u$

(c) Intensity, $I(u) = g(u)^2$

Figure 8.5: (a) Slit cross-section defines a rectangular function (Equation 8.20). (b) The function $g(u) = \operatorname{sinc}(\pi au)$ is the Fourier transform of the rectangular function in (a); it is also the field strength at the screen of light that has been diffracted by the slit (Equation 8.34). (c) Intensity $I(u) = g(u)^2$.

where $\kappa = ae^{-i2\pi r_0}$ is a constant. The sinc function is shown in Figure 8.5b. The intensity is $I(u) = g(u)^2$, which is shown in Figure 8.5c.

For example, for θ such that $\sin\theta = \lambda/a$, the field strength defined in Equation 8.33 is zero, and so the intensity in the direction θ is zero. This means that the angular width θ_W of the central peak in the diffraction pattern is given by

$$\theta_W = 2\arcsin(\lambda/a) \tag{8.36}$$
$$\approx 2\lambda/a, \tag{8.37}$$

where the approximation is justified if λ/a is small. If we choose $\lambda = 500\,\text{nm}$ (blue light) and $a = 5000\,\text{nm}$ then $\theta_W \approx 0.2$ radians, or $11.4°$.

8.4. Heisenberg's Uncertainty Principle

Quantum mechanics relies on Heisenberg's uncertainty principle, which relies on Heisenberg's inequality, which relies on Fourier analysis.

The Short Version. In the diffraction experiment described in the previous section, the wider the slit is, the more uncertain we are about the position of a photon when it passes through the slit. Accordingly, we can define the uncertainty Δx_{slit} in photon position x within the slit as the width a of the slit, i.e.

$$\Delta x_{\text{slit}} = a. \tag{8.38}$$

Similarly, we can define the uncertainty Δx_{screen} in the position on the screen as the width W of the central peak in Figure 8.5c,

$$\Delta x_{\text{screen}} = W. \tag{8.39}$$

The angular width θ_W of the central peak in the diffraction pattern is related to W by

$$\tan(\theta_W/2) = (W/2)/D, \tag{8.40}$$

and if θ_W is small then $\theta_W/2 \approx \tan(\theta_W/2) = (W/2)/D$, so that

$$\theta_W \approx W/D. \tag{8.41}$$

Putting this together with Equation 8.37 yields

$$Wa \approx 2D\lambda. \tag{8.42}$$

Substituting Equations 8.38 and 8.39, we have

$$\Delta x_{\text{screen}} \Delta x_{\text{slit}} \approx 2D\lambda. \tag{8.43}$$

Thus, given that D and λ are constant, we can reduce uncertainty in position within the slit, but only by increasing uncertainty at the screen, and vice versa. This is the essence of Heisenberg's uncertainty principle.

The Long Version. Consider a slit that is transparent at the centre but for which the transparency, and therefore intensity, falls away as a Gaussian function of distance from the centre (Figure 7.3). Accordingly, the profile of intensity within the slit is Gaussian with a standard deviation which we define to be σ_{slit}. From Section 7.11 we know that the intensity profile at the screen, being the Fourier transform of the profile at the slit, must also be Gaussian, and that the product of the standard deviation σ_{screen} of position at the screen and the standard deviation σ_{slit} of position within the slit is constant.

Here, we show that Heisenberg's uncertainty principle implies that this constant is

$$\sigma_{\text{slit}} \times \sigma_{\text{screen}} \approx \frac{D\lambda}{4\pi}, \tag{8.44}$$

where D is the slit-to-screen distance. Thus, for a given diffraction experiment, where D and λ are constant, we can trade uncertainty in position x_{slit} within the slit for uncertainty in position x_{screen} on the screen, but we cannot reduce one without increasing the other.

Conventionally, Heisenberg's uncertainty principle is expressed as the product of the standard deviation of the position (e.g. within the slit) of a particle and the standard deviation of the *momentum* of the particle. For completeness, we can work back from Equation 8.44 to the conventional statement of Heisenberg's uncertainty principle, as follows.

We assume that every particle travels in a direction parallel to the (horizontal) z-axis before reaching the slit. But as a particle exits the slit, it can travel towards the screen in direction θ (see Figure 8.3). If a particle has momentum p when it enters the slit then the conservation of momentum ensures that it still has momentum p when it exits the slit. Note that momentum has both speed and direction, so if the particle exits the slit in direction $\theta \neq 0$ then its momentum p is shared between the z and x directions.

Given that the distance between the slit and the screen is D, if a particle heads towards the screen in direction θ then it will land at

$$x_{\text{screen}} = D\tan\theta \tag{8.45}$$
$$\approx D\sin\theta, \tag{8.46}$$

where the approximation holds provided θ is small (which it is), or (equivalently) provided D is large in relation to the slit width a.

At this point, we need a brief interlude to explore a crucial observation. Given M samples of a variable x with mean $\bar{x}$, its standard deviation is

$$\sigma_x = \left(\frac{1}{M} \sum_{i=1}^{M} (x_i - \bar{x})^2 \right)^{1/2}. \tag{8.47}$$

If we now define a new variable $b = Dx$ then we find that the standard deviation of b is

$$\sigma_b = D\sigma_x. \tag{8.48}$$

With this result in mind, Equation 8.46 implies that the standard deviation of position at the screen is D times larger than the standard deviation of $\sin\theta$,

$$\sigma_{\text{screen}} \approx D\sigma_{\sin\theta}. \tag{8.49}$$

Substituting this into Equation 8.44 gives

$$\sigma_{\text{slit}} \times D\sigma_{\sin\theta} \approx \frac{D\lambda}{4\pi}, \tag{8.50}$$

and then dividing by D yields

$$\sigma_{\text{slit}} \times \sigma_{\sin\theta} \approx \frac{\lambda}{4\pi}, \tag{8.51}$$

with equality only if the distributions of x_{slit} and $\sin\theta$ are Gaussian. which is a restatement of *Heisenberg's inequality* (Equation 7.29).

We can then obtain Heisenberg's principle from a key development in quantum mechanics. Specifically, de Broglie proposed in 1924 that *any* particle has a wavelength λ that is inversely proportional to its momentum p,

$$\lambda = h/p, \tag{8.52}$$

where the constant of proportionality is *Planck's constant*, $h \approx 6.626 \times 10^{-34}$ J/Hz. Remarkably, it has been demonstrated experimentally that de Broglie's equation applies equally well to photons, electrons, atoms and even whole molecules. In other words, just as waves of light can materialise as photon particles, particles of matter form diffraction patterns when fired through a slit, as if matter can de-materialise into waves. Substituting Equation 8.52 into Equation 8.51, we get

$$\sigma_{\text{slit}} \times \sigma_{\sin\theta} = \frac{h}{4p\pi}, \tag{8.53}$$

and then multiplying both sides by p yields

$$\sigma_{\text{slit}} \times p\sigma_{\sin\theta} \approx \frac{h}{4\pi}. \qquad (8.54)$$

As mentioned above, the momentum p of a particle that is heading in direction θ can be decomposed into two orthogonal components, one of which is the component of momentum along the x-axis,

$$p_x = p\sin\theta. \qquad (8.55)$$

Using Equation 8.48, it follows that the standard deviation in momentum in the x direction is

$$\sigma_{p_x} = p\sigma_{\sin\theta}, \qquad (8.56)$$

and therefore Equation 8.54 becomes Heisenberg's uncertainty principle, expressed in terms of the uncertainty in position and in momentum,

$$\sigma_{\text{slit}} \times \sigma_{p_x} \approx \frac{h}{4\pi}. \qquad (8.57)$$

However, this equality holds only if the distributions of position x_{slit} and momentum p_x are Gaussian. This proviso leads to the conventional form of Heisenberg's uncertainty principle, expressed as the inequality

$$\sigma_{\text{slit}} \times \sigma_{p_x} \geq \frac{h}{4\pi}. \qquad (8.58)$$

Additionally, substituting $\lambda = h/p$ (Equation 8.52) into the conjugate variable $u = (\sin\theta)/\lambda$ (Equation 8.31) gives

$$u = \frac{p\sin\theta}{h}, \qquad (8.59)$$

$$= p_x/h. \qquad (8.60)$$

Substituting this and Equation 8.52 into Equation 8.30 confirms that the Fourier transform of the aperture function can be expressed in terms of momentum p_x in the x direction,

$$g(p_x) = k\int_x f(x)\, e^{-i2\pi x p_x/h}\, dx, \qquad (8.61)$$

where $k = e^{i2\pi r_0 p/h}$ (specifically, r_0 and p) is constant in each experiment.

We can check that the product of the conjugate variables x and u has no dimensions, as follows. Position x is measured in metres m, whereas $u = p_x/h$ is 1/wavelength or $1/\lambda$ (from Equation 8.52), which is therefore measured in units of $1/\text{m}$. Thus, the product of the variables $x \times p_x/h$ has no dimensions.

Finally, it is important to note that Heisenberg's uncertainty principle does not just place limits on the relative precision of *measurements* of photons at the slit and at the screen. Rather, Heisenberg's uncertainty principle places limits on the precision of what can be known, in principle, regarding photons at the slit and at the screen. Additionally, these limits apply not only to photons. As mentioned above, if electrons, atoms and even whole molecules are fired at a slit then they will form a diffraction pattern that is smaller than, but otherwise identical to, the diffraction pattern formed by light. These diffraction patterns mean that matter has wave-like behaviour, and is therefore subject to the limits imposed by Heisenberg's uncertainty principle.

8.5. Crystallography

The effects of Fraunhofer diffraction are most apparent if the wavelength of light is similar to the width of the aperture. In the case of visible light, the wavelength has a range of about 450–700 nm, and slit apertures are adjusted accordingly (e.g. 500 nm is $1/200$ of a millimetre). However, the distance between atoms in a crystal is about 0.1 nm, so if we wish to use light to observe diffraction caused by crystals then we need

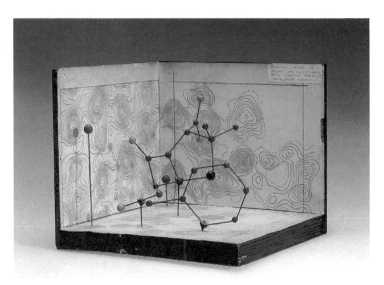

Figure 8.6: Molecular model of penicillin by Dorothy Hodgkin from 1945. The contours on each plane represent electron density within the molecule projected onto that plane. With permission from Wikimedia Commons.

electromagnetic radiation at a wavelength of about 0.1 nm. At such short wavelengths, electromagnetic radiation corresponds to X-rays.

Following pioneering work by Max von Laue around 1912, William Lawrence Bragg and his father William Henry Bragg demonstrated in 1914 that the diffraction pattern formed by exposing crystals of sodium chloride (table salt) to X-rays revealed the arrangement of atoms within the crystals. Since that time, crystallography has become an indispensable tool for finding the structure of complex biological structures, from penicillin to DNA.

Just as a slit and a screen essentially play the role of an analogue computer for producing the Fourier transform in the form of a diffraction pattern, so the distances between atoms within a crystal effectively act like atomic slits to produce a unique diffraction pattern. And just as the shape of a slit can be recovered from its diffraction pattern, so the internal structure of a crystalline substance can be recovered from its diffraction pattern. This, in essence, is crystallography.

However, a three-dimensional crystal typically has atoms arranged in planes along different orientations, where the distance between planes in each orientation gives rise to a different diffraction pattern. Consequently, it is necessary to obtain the diffraction pattern from three orthogonal axes x, y and z (e.g. by rotating the crystal). The diffraction pattern associated with each axis provides a two-dimensional Fourier transform (as in Section 7.15) of the crystal's structure, as viewed along that axis. Inverting the Fourier transform obtained from one axis provides an image that is the projection of the crystal's atomic structure along that axis. More precisely, this image is the projection of the crystal's electron density along one axis, as shown in Figure 8.6. And, because electrons cluster around atoms, a map of electron density is, for all practical purposes, a map of atomic structure.

One might think that the reliance on a crystalline structure restricts the analysis to natural crystals. However, it is possible to persuade even large biological molecules to form crystalline structures, which can then be analysed using crystallography. A historically important breakthrough for a biological molecule (penicillin) is shown in Figure 8.6, which illustrates the key ideas implicit in crystallography.

Further Reading

Beyersdorf, P (2012). An excellent brief account of Fourier optics:
https://www.youtube.com/watch?v=ofo7Urt7MQk

Douglas, B (2013). An overview of Fourier analysis (but beware of sign errors in equations). Parts 1 and 2 of this video are available at:
https://www.youtube.com/watch?v=1JnayXHhjlg
https://www.youtube.com/watch?v=kKu6JDqNma8

James, JF (2011). *A Student's Guide to Fourier Transforms.*
Brief and to the point. Some graph axes lack labels, which can be confusing for the novice; otherwise, this is excellent.

Press WH, Teukolsky SA, Vetterling WT and Flannery BP (1992). *Numerical Recipes in C. The Art of Scientific Computing.*
Well worth reading (whether or not the code is needed) for its frank writing style and practical advice about a wide variety of numerical methods (including Fourier transforms). Freely available online:
http://numerical.recipes

Riley, KF (1974). *Mathematical Methods for the Physical Sciences: An Informal Treatment for Students of Physics and Engineering.*
The best book on mathematics I have ever read, with an excellent section on Fourier transforms.

Sanderson, G. *3blue1brown.* Any video lessons by Grant Sanderson's 3blue1brown are worth watching, especially those on quantum physics:
https://www.3blue1brown.com/
Heisenberg's uncertainty principle in terms of Fourier transforms:
https://www.youtube.com/watch?v=MBnnXbOM5S4

Sivia, DS (2006). *Data Analysis: A Bayesian Tutorial.*
This superb book, which is ostensibly about Bayesian analysis, also covers many other topics, and includes a fine section on deconvolution.

Stone, JV (2015). *Information Theory: A Tutorial Introduction.*

Stone, JV (2020). *The Quantum Menagerie: A Tutorial Introduction to the Mathematics of Quantum Mechanics.*

Appendix A

A Simple Example in Python

To run this code, set the variable `testing` to 0 to analyse voice data in the supplied sound file, or set `testing` to 1 to analyse synthetic data containing just two frequencies. With `testing` set to 0 (voice data), the results are depicted in Figure 1.8. Note that this code is written for transparency, not speed. It does not use complex numbers; instead, the Fourier coefficients are found using sine and cosine functions. This code can be downloaded from https://github.com/jgvfwstone/Fourier.

```
#!/usr/bin/env python3
# -*- coding: utf-8 -*-
Fourier transform using Python 3.5 code.
############################
# Simple Fourier transform key variables
#
# x        signal of N sampled values
# T        length of signal in seconds
# R        sampling rate in samples per second
# N        number of samples in signal
# nvec     vector of N ample numbers
# tvec     vector of N time values in seconds
# fundamental  vector of N elements containing values
from 0 to 2*pi =
# fundamental frequency
# fmax  maximum frequency in Hz, chosen by user
# dt        interval in seconds between samples  = 1/R
# fmin   lowest frequency in Hz = 1/T
# nfrequencies  number of frequencies in transform
# Cs    vector of nfrequencies cosine coefficients
# Ds    vector of nfrequencies sine coefficients
# As    vector of nfrequencies Fourier amplitudes
# freqs vector of frequency values in Hz
############################
```

A A Simple Example in Python

```python
import numpy as np
import matplotlib.pyplot as plt
from scipy.io import wavfile

testing = 0 # set to 1 for synthetic data, and 0 for voice data.
fs = 10 # set fontsize for graphs.

################################################
if testing: # make simple data to test program.
    frequency = 500; # insert this frequency Hz
    T = 0.5; # length of signal in seconds

    period = 1/frequency; # in seconds
    R = 44100; # set sampling rate, samples/s
    N = R*T; # number of samples in T seconds
    tvec = np.arange(0, N, 1)/R;

    # make vector containg 1 frequency = 1/period.
    x1 = 2 * np.pi * tvec/period;

    # signal in which each interval of period seconds
    increases by 1.5*2*pi.
    x2 = 1.5*x1; # make sinusoids
    x = np.sin(x1);
    x = x + 0.5*np.sin(x2);
else: # use data from sound file
    # set name of sound file
    fname = 'sebinpubaudiIMG_1563.wav';
    # read in data from file.
    R, x = wavfile.read(fname)
    # set length of sound segment in seconds.
    segmentseconds = 0.5;
    N = int(R * segmentseconds);
    # sample number at end of segmentseconds.
    # set start and end indices in data samples.
    xmin = 0;
    xmax = xmin+N;
    x = x[xmin:xmax];
    T = N/R; # number of seconds in recording
# endif
################################################
```

```python
N = len(x); # number of samples.

# specify maximum frequency to be analysed here
fmax = 1000; # Hz

nvec = np.arange(0, N, 1);#  vector of N samples
tvec = np.arange(0, N, 1)/R;
#  vector of N samples in units of seconds

# Plot first part of signal.
nforplot=500; # number of samples for graph.
plt.figure("Signal")
plt.clf()
plt.title("Signal",fontsize=fs)
plt.xlabel("Time (seconds)",fontsize=fs)
plt.ylabel("Amplitude",fontsize=fs)
plt.plot(tvec[0:nforplot],x[0:nforplot],color="black")
plt.show()

# make fundamental sinusoidal vector.
nvec = np.arange(0,N); # indices = 0->N-1
fundamental = nvec * 2 * np.pi / N;
# fundamental now spans up to 2pi.

plt.figure("Fundamental")
plt.clf()
plt.title("Fundamental + one harmonic",fontsize=fs)
plt.xlabel("Time (seconds)",fontsize=fs)
plt.ylabel("Amplitude",fontsize=fs)
plt.plot(tvec,np.sin(fundamental))
plt.plot(tvec,np.sin(2*fundamental))
plt.show()

dt = 1/R; # interval between samples.
T = N/R;
fmin = 1.0/T; # lowest frequency in Hz.
# number of frequencies
nfrequencies = np.ceil(fmax/fmin);
nfrequencies = int(nfrequencies)
```

```python
# make arrays to store Fourier coefficients.
Cs = np.arange(nfrequencies) * 0;
Ds = np.arange(nfrequencies) * 0;
# make array to store frequency values.
freqs = np.arange(nfrequencies) * 0;

##################################################
# Here is where the work gets done.
for n in range(1, nfrequencies):

    # set frequency
    freq = fmin*n;
    freqs[n] = freq;

    # make cosine wave at frequency freq.
    cosinewave = np.cos(fundamental*n);
    # find inner product of sound with cosine wave.
    C = x.dot(cosinewave);
    Cs[n-1] = C;

    # make sine wave at frequency freq.
    sinewave = np.sin(fundamental*n);
    # find inner product of sound with sine wave.
    D = x.dot(sinewave); # equivalent to C=sum( x .* sinewave)
    Ds[n-1] = D;
# work now done.
##################################################

# find amplitude of components, normalise by N.
As = ((Cs**2 + Ds**2) ** 0.5) / N

# Plot amplitude spectrum
plt.figure("Amplitude spectrum")
plt.clf()
plt.title("Amplitude spectrum",fontsize=fs)
plt.plot(freqs,As)
plt.xlabel("Frequency (Hz)",fontsize=fs)
plt.ylabel("Amplitude",fontsize=fs)
print('\nFourier program completed.')

# END OF FILE.
```

Appendix B

A Simple Example in Matlab

For instructions, see Appendix A. This code can be downloaded from https://github.com/jgvfwstone/Fourier.

```
%%%%%%%%%%%%%%%%%%%%%%%%%%%%%%%%
% Note that this code is written for transparency, not speed.
% Simple Fourier transform key variables
%
% x       signal of N sampled values
% T       length of signal in seconds
% R       sampling rate in samples per second
% N       number of samples in signal
% tvec    vector of N time values in seconds
% nvec    vector of N elements containing
% values from 0 to 2*pi =
% fundamental frequency
% fmax    maximum frequency in Hz, chosen by user
% Nyquistfrequency Nyquist frequency= R/2
% dt          interval in seconds between samples  = 1/R
% fmin    lowest frequency in Hz = 1/T
% nfrequencies  number of frequencies in transform
% Cs     vector of nfrequencies cosine coefficients
% Ds     vector of nfrequencies sine coefficients
% As     vector of nfrequencies Fourier amplitudes
% freqs vector of frequency values in Hz
%%%%%%%%%%%%%%%%%%%%%%%%%%%%%%%

clear all;
% set testing to 0 or 1.
testing = 1;
fontsize = 15;
```

```
if testing
% make signal of two sinusoids with known frequencies
    R = 44100; % set sampling rate, samples/s
    frequency = 1000; % Hz
    period = 1/frequency; % in seconds

    T = 0.5; % length of signal in seconds
    N = R*T; % number of samples in T seconds
    tvec = 1:N; % time vector of N samples
    tvec = tvec/R; % time vector in units of seconds

    % signal in which each interval of period
    %seconds increases by 2*pi.
    x1 = 2*pi*tvec/period;

    % signal in which each interval of period
    %seconds increases by 1.5*2*pi.
    x2 = 1.5*x1;

    % make sinusoids
    x = sin(x1);
    x = x + sin(x2);

else % not testing, use singing voice as signal x.

    % set file name where signal is stored
    fname = 'sebinpubaudiIMG_1563.m4a';

    % Read in voice recording.
    [x,R] = audioread(fname); % R = samples/s

    % choose length of sound segment in seconds.
    segmentseconds = 1;
    N = R * segmentseconds;
    xmin=1; %1024*5;
    xmax=xmin+N-1;
    x=x(xmin:xmax);
    N = length(x); % number of samples.
    T = N/R;% number of seconds in recording
    % make time vector
    tvec = 1:N; % vector of sample numbers in recording.
    tvec= tvec/R; % convert to seconds.
end
```

```
figure(1);
plot(tvec,x);

% listen to sound - this may need a mac to function.
if exist('sound') sound(x,R); end

% plot sound wave.
figure(1);
plot(tvec,x,'k');
xlabel('Time (seconds)')
ylabel('Amplitude');
set(gca,'XLim',[0 T]);
% prettify graph.
set(gca,'FontSize',fontsize);
set(gca,'Linewidth',2);
hline = findobj(gcf, 'type', 'line');
set(hline,'LineWidth',2);

% make fundamental sinusoidal vector.
nvec = 1:N;
nvec = nvec*2*pi/N; % nvec now contains single wave.
nvec = nvec';

% specify maximum frequency here
fmax = 2000; % Hz

Nyquistfrequency = R/2;
if fmax > Nyquistfrequency
    error('Maximum freqency exceeds Nyquist frequency.');
end

dt = 1/R; % interval between samples.
T = N/R;
fmin = 1/T; % lowest frequency in Hz.
% number of frequencies in Fourier transform.
nfrequencies = floor(fmax/fmin);

% ensure x is a row vector
if iscolumn(x) x = x'; end
```

```
% make storage for Fourier coefficients.
Cs = zeros(1,nfrequencies);
Ds = zeros(1,nfrequencies);
% make storage for Fourier frequencies.
freqs =  zeros(1,nfrequencies);

% Here is where the work gets done.
for n = 1:nfrequencies
    % set frequency
    freq = fmin*n;
    freqs(n) = freq;

    % make cosine waves at frequency freq.
    cosinewave = cos(nvec*n);
    % find inner product of sound with cosine wave.
    % x is row vector, wave is column vector,
    % so inner product is a scalar.
    C = x * cosinewave;
    % equivalent to C=sum( x .* cosinewave)
    Cs(n) = C;

    % make sine waves at frequency freq.
    sinewave = sin(nvec*n);
    % find inner product of sound with sine wave.
    D = x * sinewave;
    Ds(n) = D;
end

% get Fourier amplitudes
As = sqrt(Cs.^2 + Ds.^2);

% get phases phasespectrum = atan(Ds./Cs);

figure(2);
plot(freqs,As,'k');
xlabel('Frequency (Hz)')
ylabel('Amplitude');
% prettify graph.
set(gca,'FontSize',fontsize);
set(gca,'Linewidth',2);
hline = findobj(gcf, 'type', 'line');
set(hline,'LineWidth',2)

% END OF FILE.
```

Appendix C

Mathematical Symbols

$*$ convolution operator, also complex conjugate (as superscript; see $\hat{A}^*$).

$|A|$ vertical bars denote a) absolute value, b) modulus (vector length), or c) magnitude of a complex number (see Glossary).

A scalar amplitude, or magnitude or modulus, of a wave.

$\hat{A}$ complex coefficient. Complex variables are written with a hat symbol.

$\hat{A}^*$ complex conjugate of $\hat{A}$. If $\hat{A} = x + iy$ then $\hat{A}^* = x - iy$.

$\boldsymbol{A}$ set of Fourier amplitudes $\{A_0/2, A_1, A_2, \dots\}$.

$\boldsymbol{C}$ set of Fourier coefficients $\{C_0/2, C_1, C_2, \dots\}$.

$\boldsymbol{D}$ set of Fourier coefficients $\{D_0/2, D_1, D_2, \dots\}$.

E energy (in units of joules, J).

e constant $2.71828\dots$.

$f(t)$ a function of time.

h Planck's constant, $\approx 6.626 \times 10^{-34}$ J/Hz.

Hz hertz, or cycles/s, unit of frequency.

i unit imaginary number, $i^2 = -1$.

J joule, unit of energy.

m metre, unit of length.

nm nanometre, unit of length equal to 10^{-9} m.

$\boldsymbol{\theta'}$ (theta) set of Fourier phase offsets $\{\theta'_0, \theta'_1, \theta'_2, \dots\}$.

s second, unit of time.

σ (sigma) standard deviation.

t time (in units of seconds).

T period, time to complete one cycle of 2π radians.

θ (theta) angle; phase of a wave, $\theta = \omega t = 2\pi\nu t$.

ω (omega) angular frequency, $\omega = 2\pi\nu$ rad/s.

ν (nu) frequency (in units of cycles per second, s^{-1} or Hz).

W watt, unit of power, 1 W $= 1$ J/s.

x position along the x-axis.

y position along the y-axis.

z complex number, $z = x + iy$; also position along the z-axis.

z^* complex conjugate, $z^* = x - iy$, of the complex number $z = x + iy$.

Appendix D

Key Equations

Consider a curve defined by a function $f(t)$ over an interval of T seconds. Angular frequency ($T = $ oscillation period):

$$\omega = 2\pi/T \text{ rad/s.} \tag{D.1}$$

Temporal frequency ($T = $ oscillation period):

$$\nu = 1/T \tag{D.2}$$
$$= \omega/(2\pi) \text{ cycles/s or Hz} \tag{D.3}$$

so

$$\omega = 2\pi\nu \text{ rad/s.} \tag{D.4}$$

The Amplitude–Phase Fourier Transform. The *amplitude–phase* form of the Fourier transform of $f(t)$ is the set of coefficient–angle pairs $\{A_n, \theta'_n\}$ such that (from Equation 2.19)

$$f(t) = A_0/2 + \sum_{n=1}^{\infty} A_n \cos(\omega_n t - \theta'_n), \tag{D.5}$$

where A_n is the amplitude of the sinusoidal Fourier component of $f(t)$ with frequency $n\omega_1$ and θ'_n specifies the phase at that frequency.

The *amplitude spectrum* is the set of amplitudes $\{A_n\}$ in the Fourier transform $\{A_n, \theta'_n\}$ of $f(t)$. If we knew the phases $\{\theta'\}$ then we could obtain the amplitude spectrum as (from Equation 4.27)

$$A_n = \frac{2}{T} \int_{t=-T/2}^{T/2} f(t) \cos(\omega_n t - \theta'_n) \, dt. \tag{D.6}$$

In practice, we do not know the phases $\{\theta'_n\}$, which can be obtained from the sine–cosine Fourier transform or the complex Fourier transform.

The Sine–Cosine Fourier Transform. The sine–cosine Fourier transform of $f(t)$ is the set of coefficients $\{C_n, D_n\}$ such that (from Equation 4.15)

$$f(t) \;=\; C_0/2 + \sum_{n=1}^{\infty} C_n \cos(\omega_n t) + D_n \sin(\omega_n t), \qquad (D.7)$$

where $C_0 = A_0$ in Equation D.5. The function $f(t)$ is the *inverse Fourier transform* of $\{C_n, D_n\}$, where (from Equations 4.24 and 4.25)

$$C_n \;=\; \frac{2}{T} \int_{t=-T/2}^{T/2} f(t) \cos(\omega_n t)\, dt, \qquad (D.8)$$

$$D_n \;=\; \frac{2}{T} \int_{t=-T/2}^{T/2} f(t) \sin(\omega_n t)\, dt. \qquad (D.9)$$

The mapping from sine–cosine coefficients to amplitude is given by

$$A_n \;=\; \sqrt{C_n^2 + D_n^2}\,, \qquad (D.10)$$

and the phase θ'_n of the sinusoidal component at frequency ω_n is given by

$$\tan \theta'_n \;=\; D_n/C_n. \qquad (D.11)$$

The Complex Fourier Transform. The Fourier transform of $f(t)$ is the set of complex coefficients $\{\hat{A}_n\}$ such that (from Equation 6.3)

$$f(t) \;=\; \sum_{n=-\infty}^{\infty} \hat{A}_n e^{i\omega_n t}, \qquad (D.12)$$

where each complex coefficient is (from Equation 6.22)

$$\hat{A}_n \;=\; \frac{1}{T} \int_{t=-T/2}^{T/2} f(t)\, e^{-i\omega_n t}\, dt. \qquad (D.13)$$

Symmetric Fourier Pairs. As described in Section 6.3, we can write the Fourier pair in a symmetric form (Equations 6.36 and 6.37)

$$f(t) \;=\; \frac{1}{\sqrt{2\pi}} \int_{\omega=-\infty}^{\infty} g(\omega)\, e^{i\omega t}\, d\omega, \qquad (D.14)$$

$$g(\omega) \;=\; \frac{1}{\sqrt{2\pi}} \int_{t=-\infty}^{\infty} f(t)\, e^{-i\omega t}\, dt. \qquad (D.15)$$

Appendix E

Glossary

aliasing When sinusoids with frequencies above the Nyquist frequency ν_{Nyq} masquerade as sinusoids with frequencies below ν_{Nyq}.

amplitude The maximum height of a wave.

amplitude spectrum The amplitude of each frequency in an ordered set of frequencies.

angular frequency For a wave with a period of T seconds, the angular frequency is $\omega = 2\pi/T$ radians per second (rad/s). See wavenumber.

conjugate variables If the Fourier transform of a function $f(x)$ of x is expressed as a sum of sinusoidal functions $g(\nu)$ of frequency ν then x and ν are conjugate variables.

complex number A number consisting of two parts, a *real part* and an *imaginary part*.

continuous Fourier transform A Fourier transform in integral form which returns coefficients that are continuous functions of frequency. Compare with discrete Fourier transform.

correlation The correlation between two zero-mean variables x and y is $(1/N)\sum_{i=1}^{N} x_i y_i$. For Gaussian distributions x and y, if they have a correlation of zero (i.e. are decorrelated) then they are also independent.

discrete Fourier transform A Fourier transform where the frequencies take discrete values, i.e. whole-number multiples of a fundamental frequency. The 'standard' discrete Fourier transform requires $O(N^2)$ operations (e.g. Appendix A), where N is the number of data samples. It can also be implemented as a fast Fourier transform.

even function A function that ignores the sign of its argument, e.g. $\cos(-x) = \cos x$. Compare with odd function.

fast Fourier transform An implementation of the discrete Fourier transform with time complexity $O(N \log N)$.

Fourier transform Almost any function $f(x)$ can be expressed as a weighted sum of sinusoids at different frequencies, where each weight is a pair of parameters that specifies the amplitude and phase of one sinusoid. Each weight can be expressed in terms of either a) the amplitude and phase, b) the amplitudes of a sine and a cosine, or c) a complex number. These weights constitute the Fourier transform of $f(x)$.

hertz (Hz) A unit of frequency, corresponding to one complete cycle of 2π radians per second.

intensity The rate at which energy is delivered to a surface, measured in watts per square metre (W/m^2).

modulus Length of a vector z, e.g. for $z = (x, y)$ it is $|z| = \sqrt{x^2 + y^2}$. Also refers to the magnitude of a complex number.

momentum In classical physics, momentum is mass $\times$ velocity, $p = Mv$ (in units of kg m/s). In quantum physics, momentum is quantised such that $p = h\nu$, where h is Planck's constant and ν is frequency.

noise Any measured physical quantity contains signal plus noise, where noise is the unwanted part of the measured quantity.

odd function A function $f(x)$ for which $f(-x) = -f(x)$, e.g. $\sin(-x) = -\sin x$. Compare with **even function**.

orthogonal Perpendicular, at $90°$ to a reference axis.

phase spectrum The phase (initial angle) associated with each frequency in an ordered set of frequencies.

power The rate of energy output, measured in J/s or W.

radian There are 2π radians in a circle, so 1 radian$= 360°/(2\pi) \approx 57.3°$.

real function A function that returns a real number.

Shannon entropy A measure of uncertainty. If x adopts values $x_1, \ldots, x_N$ with probabilities $p(x_1), \ldots, p(x_N)$ then the Shannon entropy of $p(x)$ is the average information associated with each value of x, which is $H(x) = \sum_i p(x_i) \log_2[1/p(x_i)]$ bits.

signal Any measured physical quantity contains signal plus noise. The signal is the part of the quantity that one is aiming to measure.

sinusoid A wave defined by $f(\theta) = C \cos \theta + D \sin \theta$.

spatial frequency Spatial analogue of the temporal frequency ν. For a wavelength λ, the spatial frequency is $\nu_S = 1/\lambda$ cycles/m.

standard deviation For a variable x with mean $\bar{x}$, its standard deviation is estimated as $\sigma = [(1/N) \sum_i^N (x_i - \bar{x})^2]^{1/2}$.

temporal frequency Number of cycles per second, measured in units of s^{-1}, or hertz (Hz). Given that each cycle is 2π radians, a temporal frequency of ν Hz equals an angular frequency of $\omega = 2\pi\nu$ rad/s.

theorem A mathematical statement that has been proved to be true.

variance For variable x with mean $\bar{x}$, its variance is estimated as $\sigma^2 = (1/N) \sum_i^N (x_i - \bar{x})^2$, which the square of the standard deviation.

wavelength The distance λ between consecutive peaks of a wave.

wavenumber Given a sinusoidal wave such that one wavelength of λ metres is swept out for every complete revolution of 2π radians, the wavenumber is $k = 2\pi/\lambda$ rad/m. Compare with **angular frequency**.

Index

A Note from the Author

I sincerely hope that you enjoyed reading this book. If you did (or even if you didn't) then I would be grateful if you could write a review, either on Goodreads or on Amazon.

If you think you are not sufficiently expert to write a review then you are precisely the type of reader that other readers value the most. After all, a book with the words "A Tutorial Introduction" in its title should be reviewed mainly by non-experts.

James V Stone.

Made in United States
Troutdale, OR
09/22/2023

13117159R00060